Lecture Notes in Computer Science 14732

The series Lecture Notes in Computer Science (LNCS), including its subseries Lecture Notes in Artificial Intelligence (LNAI) and Lecture Notes in Bioinformatics (LNBI), has established itself as a medium for the publication of new developments in computer science and information technology research, teaching, and education.

LNCS enjoys close cooperation with the computer science R & D community, the series counts many renowned academics among its volume editors and paper authors, and collaborates with prestigious societies. Its mission is to serve this international community by providing an invaluable service, mainly focused on the publication of conference and workshop proceedings and postproceedings. LNCS commenced publication in 1973.

Heidi Krömker
Editor

HCI in Mobility, Transport, and Automotive Systems

6th International Conference, MobiTAS 2024
Held as Part of the 26th HCI International Conference, HCII 2024
Washington, DC, USA, June 29 – July 4, 2024
Proceedings, Part I

Editor
Heidi Krömker
Technische Universitat Ilmenau
Ilmenau, Germany

ISSN 0302-9743 ISSN 1611-3349 (electronic)
Lecture Notes in Computer Science
ISBN 978-3-031-60476-8 ISBN 978-3-031-60477-5 (eBook)
https://doi.org/10.1007/978-3-031-60477-5

This Springer imprint is published by the registered company Springer Nature Switzerland AG
The registered company address is: Gewerbestrasse 11, 6330 Cham, Switzerland

Foreword

This year we celebrate 40 years since the establishment of the HCI International (HCII) Conference, which has been a hub for presenting groundbreaking research and novel ideas and collaboration for people from all over the world.

The HCII conference was founded in 1984 by Prof. Gavriel Salvendy (Purdue University, USA, Tsinghua University, P.R. China, and University of Central Florida, USA) and the first event of the series, "1st USA-Japan Conference on Human-Computer Interaction", was held in Honolulu, Hawaii, USA, 18–20 August. Since then, HCI International is held jointly with several Thematic Areas and Affiliated Conferences, with each one under the auspices of a distinguished international Program Board and under one management and one registration. Twenty-six HCI International Conferences have been organized so far (every two years until 2013, and annually thereafter).

Over the years, this conference has served as a platform for scholars, researchers, industry experts and students to exchange ideas, connect, and address challenges in the ever-evolving HCI field. Throughout these 40 years, the conference has evolved itself, adapting to new technologies and emerging trends, while staying committed to its core mission of advancing knowledge and driving change.

As we celebrate this milestone anniversary, we reflect on the contributions of its founding members and appreciate the commitment of its current and past Affiliated Conference Program Board Chairs and members. We are also thankful to all past conference attendees who have shaped this community into what it is today.

The 26th International Conference on Human-Computer Interaction, HCI International 2024 (HCII 2024), was held as a 'hybrid' event at the Washington Hilton Hotel, Washington, DC, USA, during 29 June – 4 July 2024. It incorporated the 21 thematic areas and affiliated conferences listed below.

A total of 5108 individuals from academia, research institutes, industry, and government agencies from 85 countries submitted contributions, and 1271 papers and 309 posters were included in the volumes of the proceedings that were published just before the start of the conference, these are listed below. The contributions thoroughly cover the entire field of human-computer interaction, addressing major advances in knowledge and effective use of computers in a variety of application areas. These papers provide academics, researchers, engineers, scientists, practitioners and students with state-of-the-art information on the most recent advances in HCI.

The HCI International (HCII) conference also offers the option of presenting 'Late Breaking Work', and this applies both for papers and posters, with corresponding volumes of proceedings that will be published after the conference. Full papers will be included in the 'HCII 2024 - Late Breaking Papers' volumes of the proceedings to be published in the Springer LNCS series, while 'Poster Extended Abstracts' will be included as short research papers in the 'HCII 2024 - Late Breaking Posters' volumes to be published in the Springer CCIS series.

I would like to thank the Program Board Chairs and the members of the Program Boards of all thematic areas and affiliated conferences for their contribution towards the high scientific quality and overall success of the HCI International 2024 conference. Their manifold support in terms of paper reviewing (single-blind review process, with a minimum of two reviews per submission), session organization and their willingness to act as goodwill ambassadors for the conference is most highly appreciated.

This conference would not have been possible without the continuous and unwavering support and advice of Gavriel Salvendy, founder, General Chair Emeritus, and Scientific Advisor. For his outstanding efforts, I would like to express my sincere appreciation to Abbas Moallem, Communications Chair and Editor of HCI International News.

July 2024 Constantine Stephanidis

HCI International 2024 Thematic Areas and Affiliated Conferences

- HCI: Human-Computer Interaction Thematic Area
- HIMI: Human Interface and the Management of Information Thematic Area
- EPCE: 21st International Conference on Engineering Psychology and Cognitive Ergonomics
- AC: 18th International Conference on Augmented Cognition
- UAHCI: 18th International Conference on Universal Access in Human-Computer Interaction
- CCD: 16th International Conference on Cross-Cultural Design
- SCSM: 16th International Conference on Social Computing and Social Media
- VAMR: 16th International Conference on Virtual, Augmented and Mixed Reality
- DHM: 15th International Conference on Digital Human Modeling & Applications in Health, Safety, Ergonomics & Risk Management
- DUXU: 13th International Conference on Design, User Experience and Usability
- C&C: 12th International Conference on Culture and Computing
- DAPI: 12th International Conference on Distributed, Ambient and Pervasive Interactions
- HCIBGO: 11th International Conference on HCI in Business, Government and Organizations
- LCT: 11th International Conference on Learning and Collaboration Technologies
- ITAP: 10th International Conference on Human Aspects of IT for the Aged Population
- AIS: 6th International Conference on Adaptive Instructional Systems
- HCI-CPT: 6th International Conference on HCI for Cybersecurity, Privacy and Trust
- HCI-Games: 6th International Conference on HCI in Games
- MobiTAS: 6th International Conference on HCI in Mobility, Transport and Automotive Systems
- AI-HCI: 5th International Conference on Artificial Intelligence in HCI
- MOBILE: 5th International Conference on Human-Centered Design, Operation and Evaluation of Mobile Communications

List of Conference Proceedings Volumes Appearing Before the Conference

1. LNCS 14684, Human-Computer Interaction: Part I, edited by Masaaki Kurosu and Ayako Hashizume
2. LNCS 14685, Human-Computer Interaction: Part II, edited by Masaaki Kurosu and Ayako Hashizume
3. LNCS 14686, Human-Computer Interaction: Part III, edited by Masaaki Kurosu and Ayako Hashizume
4. LNCS 14687, Human-Computer Interaction: Part IV, edited by Masaaki Kurosu and Ayako Hashizume
5. LNCS 14688, Human-Computer Interaction: Part V, edited by Masaaki Kurosu and Ayako Hashizume
6. LNCS 14689, Human Interface and the Management of Information: Part I, edited by Hirohiko Mori and Yumi Asahi
7. LNCS 14690, Human Interface and the Management of Information: Part II, edited by Hirohiko Mori and Yumi Asahi
8. LNCS 14691, Human Interface and the Management of Information: Part III, edited by Hirohiko Mori and Yumi Asahi
9. LNAI 14692, Engineering Psychology and Cognitive Ergonomics: Part I, edited by Don Harris and Wen-Chin Li
10. LNAI 14693, Engineering Psychology and Cognitive Ergonomics: Part II, edited by Don Harris and Wen-Chin Li
11. LNAI 14694, Augmented Cognition, Part I, edited by Dylan D. Schmorrow and Cali M. Fidopiastis
12. LNAI 14695, Augmented Cognition, Part II, edited by Dylan D. Schmorrow and Cali M. Fidopiastis
13. LNCS 14696, Universal Access in Human-Computer Interaction: Part I, edited by Margherita Antona and Constantine Stephanidis
14. LNCS 14697, Universal Access in Human-Computer Interaction: Part II, edited by Margherita Antona and Constantine Stephanidis
15. LNCS 14698, Universal Access in Human-Computer Interaction: Part III, edited by Margherita Antona and Constantine Stephanidis
16. LNCS 14699, Cross-Cultural Design: Part I, edited by Pei-Luen Patrick Rau
17. LNCS 14700, Cross-Cultural Design: Part II, edited by Pei-Luen Patrick Rau
18. LNCS 14701, Cross-Cultural Design: Part III, edited by Pei-Luen Patrick Rau
19. LNCS 14702, Cross-Cultural Design: Part IV, edited by Pei-Luen Patrick Rau
20. LNCS 14703, Social Computing and Social Media: Part I, edited by Adela Coman and Simona Vasilache
21. LNCS 14704, Social Computing and Social Media: Part II, edited by Adela Coman and Simona Vasilache
22. LNCS 14705, Social Computing and Social Media: Part III, edited by Adela Coman and Simona Vasilache

23. LNCS 14706, Virtual, Augmented and Mixed Reality: Part I, edited by Jessie Y. C. Chen and Gino Fragomeni
24. LNCS 14707, Virtual, Augmented and Mixed Reality: Part II, edited by Jessie Y. C. Chen and Gino Fragomeni
25. LNCS 14708, Virtual, Augmented and Mixed Reality: Part III, edited by Jessie Y. C. Chen and Gino Fragomeni
26. LNCS 14709, Digital Human Modeling and Applications in Health, Safety, Ergonomics and Risk Management: Part I, edited by Vincent G. Duffy
27. LNCS 14710, Digital Human Modeling and Applications in Health, Safety, Ergonomics and Risk Management: Part II, edited by Vincent G. Duffy
28. LNCS 14711, Digital Human Modeling and Applications in Health, Safety, Ergonomics and Risk Management: Part III, edited by Vincent G. Duffy
29. LNCS 14712, Design, User Experience, and Usability: Part I, edited by Aaron Marcus, Elizabeth Rosenzweig and Marcelo M. Soares
30. LNCS 14713, Design, User Experience, and Usability: Part II, edited by Aaron Marcus, Elizabeth Rosenzweig and Marcelo M. Soares
31. LNCS 14714, Design, User Experience, and Usability: Part III, edited by Aaron Marcus, Elizabeth Rosenzweig and Marcelo M. Soares
32. LNCS 14715, Design, User Experience, and Usability: Part IV, edited by Aaron Marcus, Elizabeth Rosenzweig and Marcelo M. Soares
33. LNCS 14716, Design, User Experience, and Usability: Part V, edited by Aaron Marcus, Elizabeth Rosenzweig and Marcelo M. Soares
34. LNCS 14717, Culture and Computing, edited by Matthias Rauterberg
35. LNCS 14718, Distributed, Ambient and Pervasive Interactions: Part I, edited by Norbert A. Streitz and Shin'ichi Konomi
36. LNCS 14719, Distributed, Ambient and Pervasive Interactions: Part II, edited by Norbert A. Streitz and Shin'ichi Konomi
37. LNCS 14720, HCI in Business, Government and Organizations: Part I, edited by Fiona Fui-Hoon Nah and Keng Leng Siau
38. LNCS 14721, HCI in Business, Government and Organizations: Part II, edited by Fiona Fui-Hoon Nah and Keng Leng Siau
39. LNCS 14722, Learning and Collaboration Technologies: Part I, edited by Panayiotis Zaphiris and Andri Ioannou
40. LNCS 14723, Learning and Collaboration Technologies: Part II, edited by Panayiotis Zaphiris and Andri Ioannou
41. LNCS 14724, Learning and Collaboration Technologies: Part III, edited by Panayiotis Zaphiris and Andri Ioannou
42. LNCS 14725, Human Aspects of IT for the Aged Population: Part I, edited by Qin Gao and Jia Zhou
43. LNCS 14726, Human Aspects of IT for the Aged Population: Part II, edited by Qin Gao and Jia Zhou
44. LNCS 14727, Adaptive Instructional System, edited by Robert A. Sottilare and Jessica Schwarz
45. LNCS 14728, HCI for Cybersecurity, Privacy and Trust: Part I, edited by Abbas Moallem
46. LNCS 14729, HCI for Cybersecurity, Privacy and Trust: Part II, edited by Abbas Moallem

47. LNCS 14730, HCI in Games: Part I, edited by Xiaowen Fang
48. LNCS 14731, HCI in Games: Part II, edited by Xiaowen Fang
49. LNCS 14732, HCI in Mobility, Transport and Automotive Systems: Part I, edited by Heidi Krömker
50. LNCS 14733, HCI in Mobility, Transport and Automotive Systems: Part II, edited by Heidi Krömker
51. LNAI 14734, Artificial Intelligence in HCI: Part I, edited by Helmut Degen and Stavroula Ntoa
52. LNAI 14735, Artificial Intelligence in HCI: Part II, edited by Helmut Degen and Stavroula Ntoa
53. LNAI 14736, Artificial Intelligence in HCI: Part III, edited by Helmut Degen and Stavroula Ntoa
54. LNCS 14737, Design, Operation and Evaluation of Mobile Communications: Part I, edited by June Wei and George Margetis
55. LNCS 14738, Design, Operation and Evaluation of Mobile Communications: Part II, edited by June Wei and George Margetis
56. CCIS 2114, HCI International 2024 Posters - Part I, edited by Constantine Stephanidis, Margherita Antona, Stavroula Ntoa and Gavriel Salvendy
57. CCIS 2115, HCI International 2024 Posters - Part II, edited by Constantine Stephanidis, Margherita Antona, Stavroula Ntoa and Gavriel Salvendy
58. CCIS 2116, HCI International 2024 Posters - Part III, edited by Constantine Stephanidis, Margherita Antona, Stavroula Ntoa and Gavriel Salvendy
59. CCIS 2117, HCI International 2024 Posters - Part IV, edited by Constantine Stephanidis, Margherita Antona, Stavroula Ntoa and Gavriel Salvendy
60. CCIS 2118, HCI International 2024 Posters - Part V, edited by Constantine Stephanidis, Margherita Antona, Stavroula Ntoa and Gavriel Salvendy
61. CCIS 2119, HCI International 2024 Posters - Part VI, edited by Constantine Stephanidis, Margherita Antona, Stavroula Ntoa and Gavriel Salvendy
62. CCIS 2120, HCI International 2024 Posters - Part VII, edited by Constantine Stephanidis, Margherita Antona, Stavroula Ntoa and Gavriel Salvendy

https://2024.hci.international/proceedings

Preface

Human-computer interaction in the highly complex field of mobility and intermodal transport leads to completely new challenges. A variety of different travelers move in different travel chains. The interplay of such different systems, such as car and bike sharing, local and long-distance public transport, and individual transport, must be adapted to the needs of travelers. Intelligent traveler information systems must be created to make it easier for travelers to plan, book, and execute an intermodal travel chain and to interact with the different systems. Innovative means of transport are developed, such as electric vehicles and autonomous vehicles. To achieve the acceptance of these systems, human-machine interaction must be completely redesigned.

The 6th International Conference on HCI in Mobility, Transport, and Automotive Systems (MobiTAS 2024), an affiliated conference of the HCI International (HCII) conference, encouraged papers from academics, researchers, industry, and professionals, on a broad range of theoretical and applied issues related to mobility, transport, and automotive systems and their applications.

For MobiTAS 2024, a key theme with which researchers were concerned was the safety and well-being of drivers. From investigating the correlation between motion sickness and driving activities to exploring the effects of advanced head-up display technologies on driver behavior, the effects of alerts, take over strategies, as well as driver behavior and performance, each contribution offers valuable insights into the complexities of human factors in driving. A considerable number of submissions focused on the multifaceted dynamics of human trust, emotion, and cognition in the context of automated driving, exploring also autonomous driving scenarios, connected vehicles and teleoperation systems, offering invaluable insights into the future of transportation. Furthermore, in this era of rapid technological advancements and evolving user needs, several papers focused on enhanced inclusivity and user experience, discussing inclusive mobility and assistive systems, independent travel for people with intellectual disabilities, cognitive load in automotive interfaces, enhanced interaction through gestures, as well as user experience design and usability testing approaches. Finally, the design of urban mobility and public transportation systems has also collected diverse contributions on user needs and perspectives regarding public transportation, as well as human-centered design and user acceptance of urban mobility approaches, contributing to the ongoing dialogue for innovations in public transportation.

Two volumes of the HCII 2024 proceedings are dedicated to this year's edition of the MobiTAS conference. The first focuses on topics related to Driver Behavior and Safety, and Human Factors in Automated Vehicles, while the second focuses on topics related to Urban Mobility and Public Transportation, and User Experience and Inclusivity in MobiTAS.

The papers in these volumes were accepted for publication after a minimum of two single-blind reviews from the members of the MobiTAS Program Board or, in some cases, from members of the Program Boards of other affiliated conferences. I would like to thank all of them for their invaluable contribution, support, and efforts.

July 2024 Heidi Krömker

6th International Conference on HCI in Mobility, Transport and Automotive Systems (MobiTAS 2024)

The full list with the Program Board Chairs and the members of the Program Boards of all thematic areas and affiliated conferences of HCII 2024 is available online at:

http://www.hci.international/board-members-2024.php

HCI International 2025 Conference

The 27th International Conference on Human-Computer Interaction, HCI International 2025, will be held jointly with the affiliated conferences at the Swedish Exhibition & Congress Centre and Gothia Towers Hotel, Gothenburg, Sweden, June 22–27, 2025. It will cover a broad spectrum of themes related to Human-Computer Interaction, including theoretical issues, methods, tools, processes, and case studies in HCI design, as well as novel interaction techniques, interfaces, and applications. The proceedings will be published by Springer. More information will become available on the conference website: https://2025.hci.international/.

General Chair
Prof. Constantine Stephanidis
University of Crete and ICS-FORTH
Heraklion, Crete, Greece
Email: general_chair@2025.hci.international

https://2025.hci.international/

Contents – Part I

Contents – Part II

Driver Behavior and Safety

Activities that Correlate with Motion Sickness in Driving Cars – An International Online Survey

Frederik Diederichs[1(✉)], Amina Herrmanns[1], David Lerch[1], Zeyun Zhong[5], Daniela Piechnik[2,3], Lesley-Ann Mathis[3], Boyu Xian[2,3], Nicklas Vaupel[1], Ajona Vijayakumar[1], Canmert Cabaroglu[1], and Jessica Rausch[4]

[1] Fraunhofer IOSB, 76131 Karlsruhe, Germany
frederik.diederichs@iosb.fraunhofer.de
[2] University of Stuttgart IAT, 70569 Stuttgart, Germany
[3] Fraunhofer IAO, 70569 Stuttgart, Germany
[4] Ford Research Centre, 52072 Aachen, Germany
[5] Karlsruher Institute of Technology, 76131 Karlsruhe, Germany

Abstract. Up to 2 out of 3 passengers suffer from motion sickness, caused by non-driving related activities. Occupant monitoring systems detect such activities via cameras in the vehicle interior and hence can be used to warn passengers or to assist them. An international online survey in Germany, USA, China, India, Turkey and Mexico was conducted in order to identify activities that correlate with motion sickness. The results identify reading, using a device, watching a movie and turning in the seat to be the most relevant activities for occupant monitoring systems to detect and hence for motion sickness assistance systems to address.

Keywords: motion sickness · occupant monitoring system · automated driving

1 Introduction

1.1 Motion Sickness is a Common Problem in Driving Cars

Motion sickness in driving cars is a common problem for passengers, among them children and a major part of adult population. According to Reason and Brand (1975), up to 2 out of 3 passengers suffer from motion sickness. This result is confirmed in a similar relation by Schmidt et al. (2020) and by Brietzke et al. (2022).

Motion sickness, also called car sickness, is attributed to a discrepancy in the sensory signals received by the brain. There are two main theories that explain why people get motion sick (Bos et al., 2008). The Sensory Conflict Theory suggests that motion sickness occurs when there is a disconnect or conflict between the sensory inputs received by the brain. For example, when someone is reading a book in a moving car, the inner ears, which help control balance and detect motion, sense that the body is moving, but the eyes - focused on the stationary book - do not. This discrepancy in sensory information can

H. Krömker (Ed.): HCII 2024, LNCS 14732, pp. 3–12, 2024.
https://doi.org/10.1007/978-3-031-60477-5_1

lead to symptoms of motion sickness, as the brain struggles to reconcile the conflicting signals.

The Postural Instability Theory proposes that motion sickness is related to the body's inability to maintain postural stability and control in a moving environment. When the motion of the vehicle or mode of transport constantly challenges the body's balance and stability mechanisms, it can lead to the development of motion sickness symptoms.

Both theories highlight the importance of consistent and congruent sensory inputs for maintaining a sense of equilibrium and preventing the discomfort associated with motion sickness. Hence, motion sickness while driving develops mainly during (repeated) accelerations of the driving car, either in lane changes, curves and roundabouts or in stop and go situations. It is often related to activities where the eyes are not looking outside of the car.

1.2 Motion Sickness is a Problem for Automated Driving

With the introduction of automated driving of Level 3 and 4 (SAE J3016, 2021) motion sickness can be a problem for the acceptance of automated vehicles. While automated driving can make driving more productive and comfortable, motion sickness can negate these positive aspects. The symptoms of motion sickness reduce wellbeing and driving capability. Hence, research into the prevention of motion sickness is a relevant research topic towards the introduction of automated cars. Prevention, detection and countermeasures are under investigation in several studies (Diels et al., 2016, Brietzke et al., 2017, Hainsch et al., 2021, Bos et al., 2022).

People who develop motion sickness, suffer from very unpleasant symptoms of sweating, pallor, nausea and vomiting (Diels et al. 2016). The intensity of the symptoms varies a lot. Symptoms can last between several minutes and up to hours. People who perceive medium or high degrees of motion sickness feel a strong decrease of judgment capability, reduced environmental perception capabilities and slower reaction times (Bos et al., 2008, Smyth et al., 2018).

1.3 Motion Sickness Correlates with Non-driving Related Activities

The development of motion sickness in driving cars correlates with certain non-driving related activities (NDRA). People who tend to get motion sick, avoid such activities. Hence, they have a handicap in freely using their time in a driving car.

Schmidt et al. identified the following NDRA in a large international online survey, which we have adopted in this survey: Reading, using devices, watching videos, looking out the window, turning backwards, sleeping and looking out the window are NDRA that car passengers relate to motion sickness (Schmidt et al., 2020, Brietzke et al., 2022).

1.4 Activity Detection Can Be Used to Assist Non-driving Passengers

Knowing activities that correlate with motion sickness can be used as additional information in route navigation devices. Routes or stretches of routes can be classified to be suitable for certain activities.

Users who are not driving themselves and are engaged in NDRA could also be alerted ahead of winding stretches of road and stop and go situations. This approach is investigated in the KARLI project (Diederichs et al., 2022). The authors know from not published interviews that people sometimes forget about their motion sickness risk or voluntarily take the risk of getting motion sick, in order to get something done. Sometimes they realize too late that motion sickness has developed. Adults reported that for this reason they would welcome reminders of motion sickness risk. Parents mentioned that they provide this assistance to kids frequently.

Intelligent user interfaces could also refer to the NDRA in notifications, thereby increasing the acceptance of the notifications. For such notifications with high user acceptance, it is necessary to know, among other things, which NDRA are particularly associated with the development of motion sickness. Detection systems can then be developed and optimized to detect these NDRA. There is currently no empirical information available on the most important activities for preventing motion sickness.

1.5 Occupant Monitoring Systems Detect Activities

Occupant monitoring systems consist of visual sensors such as RGB, IR or depth cameras and algorithms processing the visual information. These algorithms are mostly machine learning (ML) based. In order to train ML algorithms, a huge amount of data is necessary. Currently only few datasets for in-cabin monitoring are publicly available, such as Drive&Act (Martin et al., 2019) and Driver Monitoring Dataset (Ortega et al. 2020). Current occupant monitoring systems (Martin et al., 2019, Zhong et al., 2023) detect various secondary activities via cameras. Data collection and labeling is time consuming and expensive, therefore techniques like synthetic data or generative models (Lerch et al. 2024) can be used to generate training data.

1.6 Which Activities Correlate with Motion Sickness?

There is currently no measure available to detect motion sickness in cars, consequently there are no empirical data about the occurrence of motion sickness while driving, that could be correlated with measured activities. Furthermore, getting test participants motion sick in a driving car, would require special safety measures, e.g. the use of a Wizard-of-Oz Vehicle (Diederichs et al., 2021, Piechnik et al., 2023). Asking users appears to be the best method to identify relevant activities that correlate with motion sickness.

2 Method

An international online survey was designed in order to ask users for activities that correlate with motion sickness according to their own experience. Activities found by Schmidt et al. (2020) were added to the questionnaire together with three activities that occupant monitoring systems, such as described in Martin et al. (2019) detect already: doing make-up, shaving and taking photographs. Additionally, we asked for other activities that are related with the development of motion sickness and also ask for workarounds how users perform such activities avoiding the development of motion sickness.

The questionnaire was developed in the program Qualtrics (see Fig. 1). It was available via a special website link for each country. The questionnaire was always translated by DeepL and counterchecked by a native speaker into the main language of the target country.

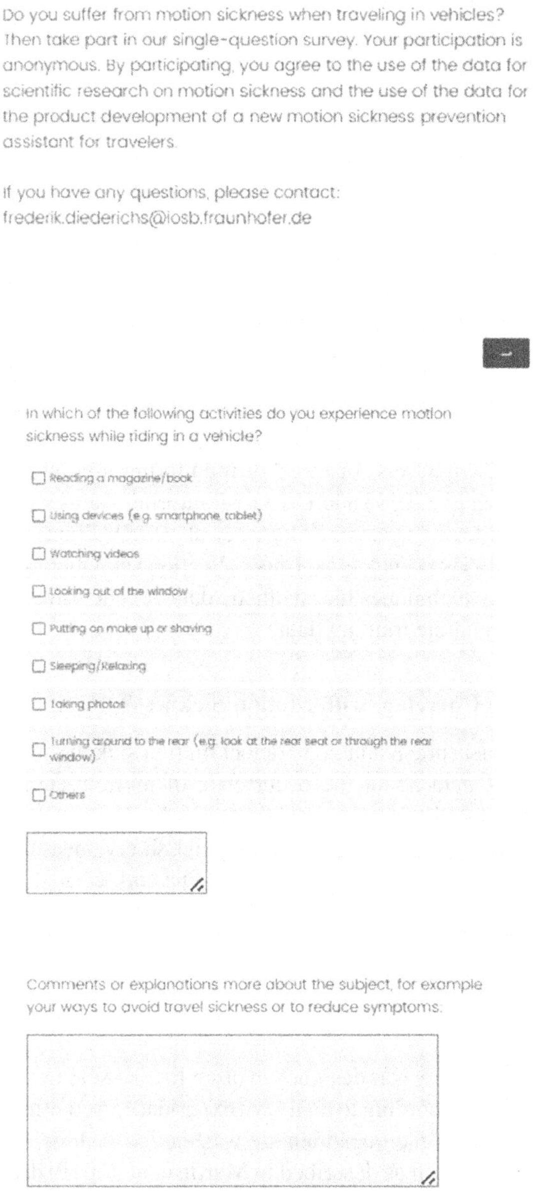

Do you suffer from motion sickness when traveling in vehicles? Then take part in our single-question survey. Your participation is anonymous. By participating, you agree to the use of the data for scientific research on motion sickness and the use of the data for the product development of a new motion sickness prevention assistant for travelers.

If you have any questions, please contact: frederik.diederichs@iosb.fraunhofer.de

In which of the following activities do you experience motion sickness while riding in a vehicle?

- ☐ Reading a magazine/book
- ☐ Using devices (e.g. smartphone, tablet)
- ☐ Watching videos
- ☐ Looking out of the window
- ☐ Putting on make up or shaving
- ☐ Sleeping/Relaxing
- ☐ Taking photos
- ☐ Turning around to the rear (e.g. look at the rear seat or through the rear window)
- ☐ Others

Comments or explanations more about the subject, for example your ways to avoid travel sickness or to reduce symptoms:

Fig. 1. Screenshot of the English version of the questionnaire about activities that correlate with motion sickness in driving cars.

Participants were recruited by the authors via WhatsApp groups and e-mails among friends, family and chat rooms in several countries. Only results with at least 50 participants per country were included in the results. It was communicated, but not controlled, that participants had experience with motion sickness either themselves or with other passengers, that participants live in the respective country and that participation was allowed only once. In order to achieve more answers, the questionnaire was very short (<2 min) and anonymous to avoid any GDPR extra efforts by the participants. Hence, no questions for age, gender, driving license or others was collected.

The questionnaire consisted of two questions. In the first question, participants could indicate (with multiple choices) which of the activities make them feel motion sick in driving cars. An open question allowed participants to add additional activities. The second question is an open question where participants were asked to name methods that help them to avoid or get rid of motion sickness.

3 Results

The voting is displayed in Table 1 and in Fig. 2 for the activities 1 to 8:

1. Reading
2. Using devices
3. Watching videos
4. Turning around to the rear
5. Looking out of the window
6. Putting on make-up or shaving
7. Taking photos
8. Sleeping/Relaxing

Table 1. The votings in numbers.

Country	N	1	2	3	4	5	6	7	8
China	**101**	37	47	32	17	18	14	7	9
USA	**51**	37	30	19	14	3	1	3	1
Mexico	**100**	50	34	15	11	16	1	5	3
India	**53**	13	14	9	12	8	4	2	4
Turkey	**60**	27	27	19	13	1	1	3	2
Germany	**132**	113	113	91	65	13	24	28	6

As countermeasures against the development of motion sickness or reliefs in case of motion sickness, participants mentioned multiple measures that can be clustered in 7 categories (see Table 2).

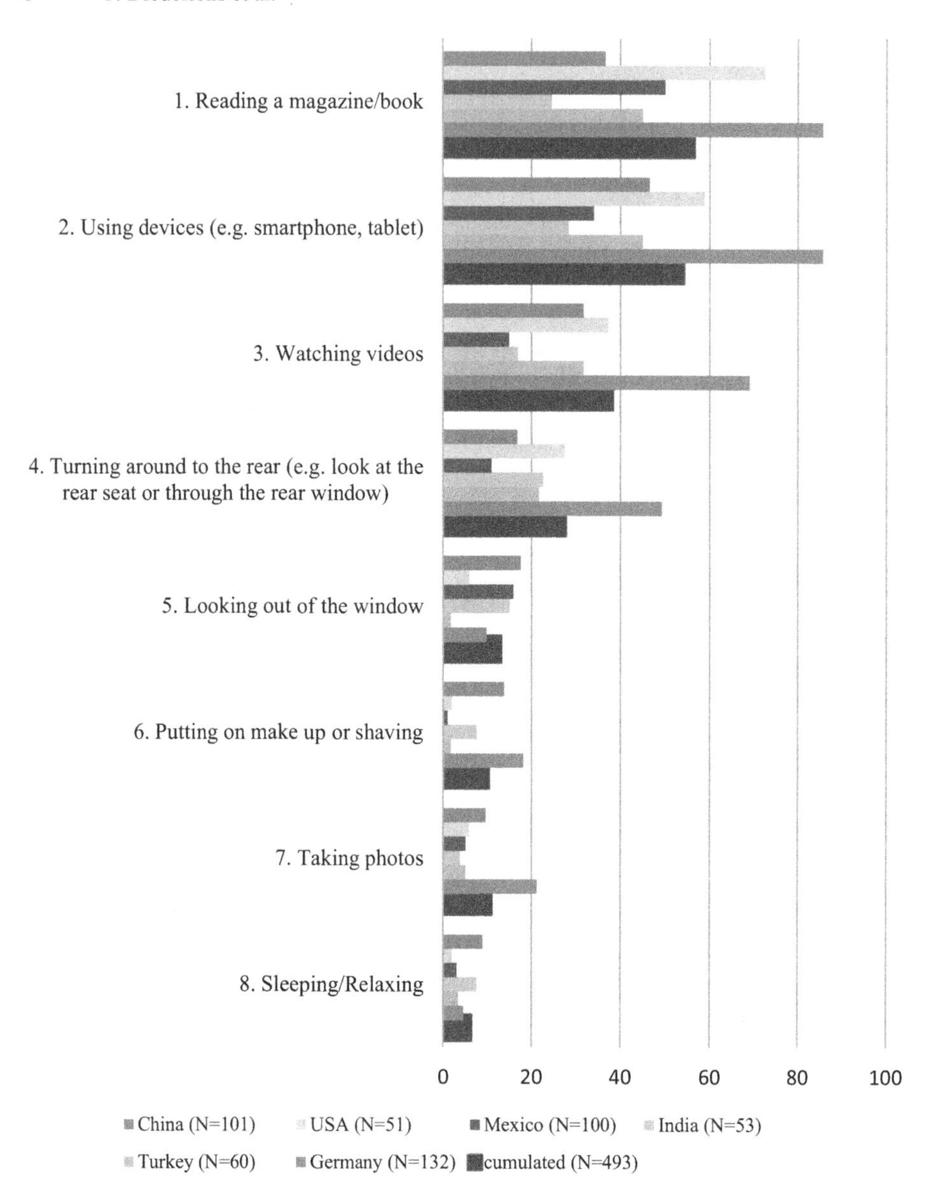

Fig. 2. Activities that correlate with motion sickness (answers in % of the total votings).

4 Interpretation

The voting results from 6 countries reveal only small differences between the countries. A cumulation of the answers from all countries seems to represent the data well.

Table 2. Countermeasures against motion sickness.

Nutrition	Drink water	
	Do not eat greasy food	
	Chew gum	
	Candies (mint or ginger flavor)	
	Acidic food (smelling)	Oranges & orange peel
		Lemons
		Ginger
	Eat nothing at all	
	No carbon dioxide	
	Vinegar-water	
Medication for motion sickness		
Sleeping/Relaxing		
Window	Look out of the window	
	Open the window/ Ventilate	
Organisation	Do not use devices/do not read	
	Only use/read devices briefly	
	Use devices/read when car is standing still	
	Hold devices/reading material at eye level	
	Keep head straight	
	Sit upright	
	Brake with foresight	
	Drive slowly	
	Sit in the front (passenger seat)	
	Low temperature in the car	
	Take breaks and get out and walk a bit	
Alternatives	Listening to music (headphones)	
	Audiobook instead of reading	
	Talking to fellow travelers	
	Talking on the phone	
Somatics	Pressing your hand against your stomach	
	Bring your body/head close to the car	
	Hegu/acupuncture points	
	Apply Windex to temples or Fenchi points	

In Germany more people selected multiple activities than in other countries. However, this does not change the percentual distribution.

Activities that more than 25% of people correlate with motion sickness are.

1. Reading
2. Using devices
3. Watching Videos
4. Turning around to the rear

The participants most frequently stated that it would help them if they stopped reading or using the devices, held the devices at window height and looked outside in between. Looking out of the window or at the street, chewing gum, sleeping or taking pills for motion sickness were also frequently mentioned.

5 Discussion and Limitations

The method of asking people about their behavior incorporates some limitations. However, in lack of data from observations this method delivered reasonable data. This data will also help to develop assistant systems with a high ecological validity in the perception of users.

The fact that users in all countries and all regions of the world seem to correlate similar activities with motion sickness may be caused by the fact, that all of them use very similar vehicles. However, a limitation of the online survey might have been, that the activities were presented in a certain order and the first activities received most of the votes. An effect of sequence might be visible in the results. Nevertheless, the last option "turning around in the seat" was moved from the last place in the survey to the fourth place in the result. The effect of order is probably not very strong. Still, the anonymous survey may have attracted people who just try to click fast through the survey – and those people may have clicked one of the first options more often.

German participants selected very often multiple activities, while participants from other countries selected less or only one activity. The translation may have had an impact on this, even though the instruction always asked for activities in plural.

The approach to use friends and family members of the researchers as multiplicators for the survey may have biased the results. However, it assured a strong dedication among participants to answer the questionnaire correctly. The anonymous participation holds a strong risk for having many votes from unmotivated people. Our approach reduced this effect probably. Still we were limited in the sample size by this approach. Depending on the contacts in the respective countries we achieved between 50 and over 100 participants per country. The differences between the activities are clear, so apparently the amount of almost 500 participants in total was sufficient to discriminate between relevant and less relevant activities.

6 Conclusion

The results show that there are selected activities that many people correlate with motion sickness in driving cars, like reading, using devices, watching videos and turning around, according to the self-reporting of almost 500 participants.

There are no relevant differences in the countries.

Occupant monitoring systems detect the relevant activities already now in research applications and can be further improved in the accuracy for especially those relevant activities: reading, using a device, watching videos and turning to the rear. With this technology and knowledge from this paper, effective assistants can be developed to support vehicle occupants, based on activity recognition.

Acknowledgments. The work has been funded under the funding codes 19A21031E, 19A21031F and 19A21031L by the Federal Ministry for Economic Affairs and Climate Action of Germany (BMWK) on the basis of a decision by the German Bundestag and by the European Union. The work was performed in the project KARLI (www.karli-projekt.de).

Disclosure of Interests. The authors have no competing interests to declare that are relevant to the content of this article, besides the affiliation that they are working for and that is disclosed in the header of the article.

References

Bos, J.E., Hogervorst, M.A., Munnoch, K., Perrault, D.: Human performance at sea assessed by dynamic visual acuity. In: ABCD Symposium' Human Performance159 Automotive UI '22 Adjunct, 17–20 September 2022, Seoul, Republic of Korea (2022)

Bos, J.E., Bles, W., Groen, E.L.: A theory on visually induced motion sickness. Displays **29**(2), 47–57 (2008)

Brietzke, A., Kantusch, T., Xuan, R.P., Dettmann, A., Marker, S., Bullinger, A.C.: What can user typologies tell us about car-sickness criticality in future mobility systems. SAE Int. J. Connected Autom. Veh. **5**(12-05-02-0012), 135–145 (2022)

Brietzke, A., Klamroth, A., Dettmann, A., Bullinger, A.C.: Motion sickness in cars: influencing human factors as an outlook towards highly automated driving. Headache **41**(14.29), 8–33 (2017)

Diederichs, F., et al.: Adaptive transitions for automation in cars, trucks, buses and motorcycles. IET Intel. Transp. Syst. **14**(8), 889–899 (2020)

Diederichs, F., Mathis, L.A., Bopp-Bertenbreiter, V., Bednorz, B., Widlroither, H., Flemisch, F.: A Wizard-of-Oz vehicle to investigate human interaction with AI-driven automated cars (2021)

Diederichs, F., et al.: Artificial intelligence for adaptive, responsive, and level-compliant interaction in the vehicle of the future (KARLI). In: Stephanidis, C., Antona, M., Ntoa, S. (eds.) HCII 2022. CCIS, vol. 1583, pp. 164–171. Springer, Cham (2022). https://doi.org/10.1007/978-3-031-06394-7_23

Diels, C., Bos, J.: Self-driving carsickness. Appl. Ergon. **53**(2016), 374–382 (2016)

Hainich, R., Drewitz, U., Ihme, K., Lauermann, J., Niedling, M., Oehl, M.: Evaluation of a human–machine interface for motion sickness mitigation utilizing anticipatory ambient light cues in a realistic automated driving setting. Information **12**(4), 176 (2021)

Lerch, D., Zhong, Z., Martin, M., Voit, M., Beyerer, J.: Unsupervised 3D skeleton-based action recognition using cross-attention with conditioned generation capabilities. In: Proceedings of the IEEE/CVF Winter Conference on Applications of Computer Vision (WACV) Workshops, pp. 211–220 (2024)

Martin, M., et al.: Drive&Act: a multi-modal dataset for fine-grained driver behavior recognition in autonomous vehicles. In: Proceedings of the IEEE/CVF International Conference on Computer Vision, pp. 2801–2810 (2019)

Ortega, J.D., et al.: DMD: a large-scale multi-modal driver monitoring dataset for attention and alertness analysis. In: Bartoli, A., Fusiello, A. (eds.) Computer Vision – ECCV 2020 Workshops. LNCS, vol. 12538, pp. 387–405. Springer, Cham (2020). https://doi.org/10.1007/978-3-030-66823-5_23

Piechnik, D., Mathis, L.-A., Diederichs, F., Lerch, D., Martin, M., Widlroither, H.: Technical setup of a Wizard-of-Oz vehicle for on-road AI data collection. In: 15th ITS European Congress Lisbon, Portugal, Lisbon, Portugal (2023)

Reason, J.T., Brand, J.J.: Motion Sickness. Academic Press (1975)

SAE Taxonomy and Definitions for Terms Related to Driving Automation Systems for On-Road Motor Vehicles J3016_202104 (2021). https://www.sae.org/standards/content/j3016_202104/

Schmidt, E.A., Kuiper, O.X., Wolter, S., Diels, C., Bos, J.E.: An international survey on the incidence and modulating factors of carsickness. Transport. Res. F Traffic Psychol. Behav. **71**, 76–87 (2020)

Smyth, J., Jennings, P., Mouzakitis, A., Birrell, S.: Too sick to drive: how motion sickness severity impacts human performance. In: 2018 21st International Conference on Intelligent Transportation Systems (ITSC), pp. 1787–1793. IEEE November 2018

Zhong, Z., Martin, M., Voit, M., Gall, J., Beyerer, J.: A survey on deep learning techniques for action anticipation (2023). arXixsv preprint arXiv:2309.17257

Drowsiness and Emotion Detection of Drivers for Improved Road Safety

Nishat Anjum Lea[1](✉), Sadia Sharmin[1], and Awal Ahmed Fime[2]

[1] Bangladesh University of Engineering and Technology, Dhaka, Bangladesh
nishatanjumlea2022@gmail.com, sadiasharmin@cse.buet.ac.bd
[2] Kent State University, Ohio, USA

Abstract. Often resulting in fatalities and serious injuries, drowsy driving is a key contributing factor in collisions. Drivers who are irritable or disturbed are more likely to be involved in accidents. Identifying symptoms of exhaustion and negative emotions may facilitate taking preventative measures before an accident occurs. By identifying and responding to the driver's state, cars equipped with sophisticated technology such as drowsiness and emotion detection aim to enhance the driving experience. With the use of machine learning algorithms, these cars can accurately read facial expressions to determine the driver's drowsiness and emotion. The technology may alter the vehicle's behaviour and user interface in response to the driver's state, making driving safer and more personalised. To detect driver's drowsiness different approaches are taken in this experiment such as the use of Basic CNN, ResNet50, VGG16, VGG19 and InceptionV3. Among them InceptionV3 has given 100% accuracy in the process of detecting drowsy drivers for yawn_eye_dataset_new. In case of NITYMED dataset where video data were used, Resnet50 produced 100% accuracy. We have used InceptionV3 for detecting driver's emotion for FER2013 Dataset, which has given 68.02% accuracy. We conducted a survey on drivers regarding what affects them at the time of driving and whether a continuous monitoring system will help them or not. Almost everyone agreed on having a monitoring system for road safety.

Keywords: Driver Drowsiness · CNN · Emotion Analysis

1 Introduction

In an era marked by technological breakthroughs and increased demands on our roads, the safety of both drivers and pedestrians has become a top priority. Driver sleepiness is a key factor to road accidents, since it impairs an individual's ability to remain aware and attentive behind the wheel. Drowsy driving has serious effects, including an increased chance of accidents and injury.

Recognising the essential role that driver attentiveness plays in road safety, there has been a rising emphasis on creating sleepiness detection technology.

H. Krömker (Ed.): HCII 2024, LNCS 14732, pp. 13–26, 2024.
https://doi.org/10.1007/978-3-031-60477-5_2

These technologies attempt to detect indicators of driver weariness and respond quickly, therefore reducing the hazards connected with impaired driving.

This study's main goal is to increase road safety by reducing accidents brought on by fatigued or upset drivers. Through study, we may create and improve technologies that accurately identify these states and put them into practise with treatments to lessen their negative effects on driving performance. Our un- derstanding of the intricate interaction between tiredness and emotions and how they influence driving behaviour will be improved through research. Improved treatments and defences may result from this understanding. The main contributions of this research are:

- From our experiment, it is proved that different CNN models work well for image data. We conducted this experiment on "yawn_eye_data_set_new" [1] and found 100% accuracy.
- The experiment was also conducted on NITYMED video dataset [2]. In this case, several images were selected from video data with full face and classified for drowsiness detection.
- We experimented with FER2013 [3] emotion data so that apart from drowsy state of driver's, their emotions can also be detected.
- We conducted a survey on drivers regarding what affects them most at the time of driving, what is the main reason of accident and whether they need a continuous monitoring system or not.

The major and novel contribution of this research is the survey that conducted on the drivers who drive different types of vehicles for personal and professional purposes. Their experiences and the issues they face while driving have been documented for future work. The survey result will help to make an user friendly technology for road safety.

2 Background

In order to find a way to prevent traffic accidents, several researchers are actively interested. H. Varun Chand et al. (2022) proposed CNN Based Driver Drowsiness Detection System Using Emotion Analysis [4]. The emotion analysis, in this proposed model, analyzes the driver's frame of mind which identifies the motivating factors for different driving patterns. The facial pattern of the driver is treated with 2D Convolution Neural Network (CNN) to detect the behavior and driver's emotion. Samy Abd El-Nabi et al. (2023) proposes a system which reviews AI techniques for driver drowsiness detection, including facial expressions, biological signals, and vehicle behavior analysis [5]. Fouad et al. proposed a robust and efficient EEG-based drowsiness detection system using different machine learning algorithms. [6] His paper addresses the issue of drowsy driving-related accidents by proposing a software-based fatigue detection system using

EEG signals. The study evaluates various machine learning algorithms, including Naive Bayes, Support Vector Machines, K-Nearest Neighbor, and Random Forest Analysis, achieving hundred percent accuracy rate for all subjects with multiple classifiers. The developed EEG-based system offers real-time detection of driver drowsiness and loss of focus, demonstrating practicality and reliability for real-time applications in enhancing road safety. Ruben Florez et al. proposed A CNN-Based approach for driver drowsiness detection by real-time eye state identification [7]. Based on InceptionV3, VGG16, and ResNet50V2, three neural networks that implement deep learning were examined. NITYMED is the database that is used; it includes recordings of drivers at varying states of drowsy. The study's findings demonstrate the excellent accuracy with which all three convolutional neural networks can identify drowsiness in the ocular region. With an average accuracy rate of 99.71%, the Resnet50V2 network in particular attained the highest accuracy rate. Mohammed Imran Basheer Ahmed et al. proposed A deep-Learning approach to driver drowsiness detection [8]. The suggested model proposes a way to use a convolutional neural network (CNN) to measure changes in a driver's ocular movement in order to assess the degree of driver fatigue, thereby addressing the problem of road safety. An architecture known as deep drowsiness detection (DDD) is shown in the research of Park et al. [9]. It analyses RGB movies that centre on the driver's complete face. Three architectures are used by the DDD architecture: FlowIma-geNet, VGG-FaceNet, and AlexNet. The output of these networks is combined to categorise the degree of drowsiness in the input video frames. The NTHU-DDD database is used by the authors to test their suggested model, and they find that on average, their experimental results yield an accuracy of 73.06%. An architecture that makes use of the ocular region especially was proposed by Chirra et al. [10]. The Viola Jones-proposed Haar Cascade method was applied to extract the ocular region. In order to identify the face and eyes, the CNN uses the ROI of the eyes as input. It then uses a database that was gathered throughout network training to achieve an accuracy of 98% during training, 97% during validation, and 96.42% during the final test. The authors of Zhao et al.'s approach [11] classified and detected tiredness using face characteristic points. For face detection and characteristic point location, they employed an MTCNN (multi-task cascaded convolutional network). They extracted ROIs from the mouth and eyes, which were then passed to their EM-CNN network. There, four classes were classified-two for the state of the eyes and two for the state of the mouth. Comparing their results to different types of designs, they achieved 93.623% accuracy using a database that the business Biteda provided for their experiments.

3 Methodology

We used different CNN Architectures for detection of drowsiness and emotion. We conducted a survey on drivers regarding what affects them and the necessity of a monitoring system. Based on their feedback we will improve our monitoring system. Currently we have designed the monitoring system using raspberry pi. The flow diagram of the whole system is shown by Fig. 1

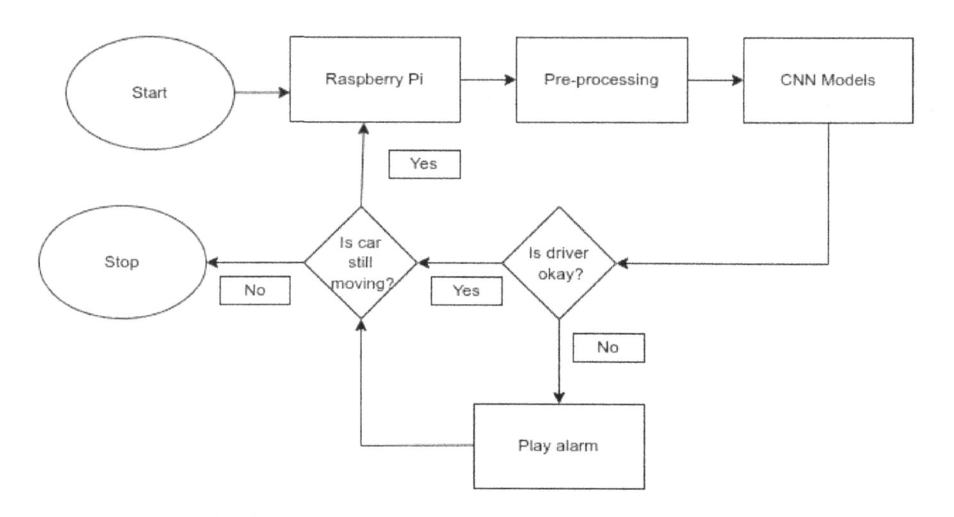

Fig. 1. Flow diagram for the whole system.

3.1 CNN Architecture

Convolutional Neural Networks (CNNs) are a type of deep learning architecture that was built primarily for image processing and recognition applications. CNNs use a hierarchical structure to learn and extract information from input pictures using convolutional layers, pooling layers, and fully connected layers. This architecture has demonstrated exceptional performance in applications like as picture categorization, object identification, and facial recognition.

ResNet (Residual Networks). Kaiming He et al. introduced ResNet to address the challenge of training very deep networks. The usage of residual blocks, which feature shortcut connections that allow the network to learn residual functions, is its main innovation. This design notably accelerates deep network training by addressing difficulties such as disappearing gradients. ResNet has performed admirably in image recognition tasks, winning the ILSVRC in 2015.

VGG16 and VGG19. VGG16 and VGG19 designs were suggested by the Visual Geometry Group (VGG) at the University of Oxford. These models are distinguished by their simplicity, since they employ a consistent architecture with modest 3×3 convolutional filters and deep layer stacking. VGG16 has 16 layers, whereas VGG19 has 19 layers. VGG networks' simple architecture enables comprehension and transferability to a variety of image recognition problems.

EfficientNet. Mingxing Tan and Quoc V. Le's EfficientNet focuses on balancing model accuracy and processing economy. To achieve optimal performance, it provides a compound scaling mechanism that uniformly scales the network's depth, breadth, and resolution. This method enables EfficientNet to obtain cutting-edge results with many fewer parameters, making it computationally more efficient than many competitors.

InceptionNet. The notion of inception modules was presented by InceptionNet, commonly known as GoogLeNet. These modules use numerous filters with varied receptive field widths in tandem to capture a variety of characteristics at various scales. InceptionV3 is a newer version that improves on the original design in terms of accuracy and performance.

3.2 Dataset

Yawn_eye_dataset_new Dataset. "yawn_eye_dataset_new" is used for the experiment. [1] There are mainly four types of data: Closed, Open, No Yawn, Yawn. There are almost 2800 image data for training with minimum 625 data for each type. For testing each category has minimum 100 data. In total 400 data for testing. Some example image data are shown below by Fig. 2, Fig. 3, Fig. 4, Fig. 5

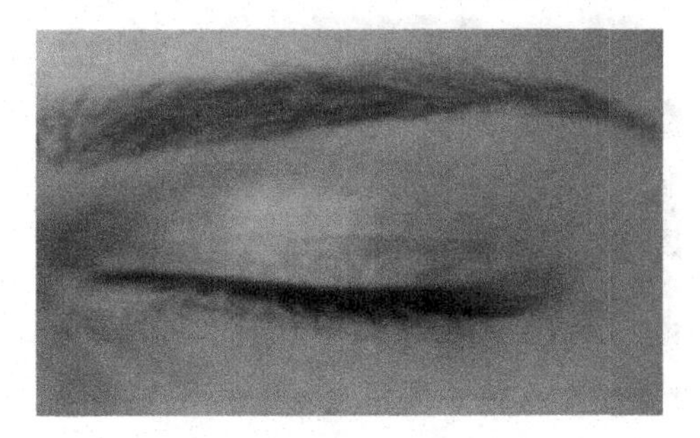

Fig. 2. Closed Eye

NITYMED Dataset. Nitymed [2] is a video dataset to detect whether the driver is yawning or not. NITYMED stands for Nighttime-Yawning-Microsleep-Eyeblink-Distraction. Using this dataset, yawning and blinking sleepy eyes have been identified. Nevertheless, this dataset can also be used to evaluate various face, mouth, and eye tracking applications (driver distraction/microsleep, facial expressions, etc.). It consists of 130 videos, among those videos 107 videos contain drivers with yawning. These videos are approximately 15–25 seconds long. These dataset is created with 21 participants (11 males and 10 females) in Patras, Greece. Face, mouth, and eyes of the drivers are visible while driving a real car.

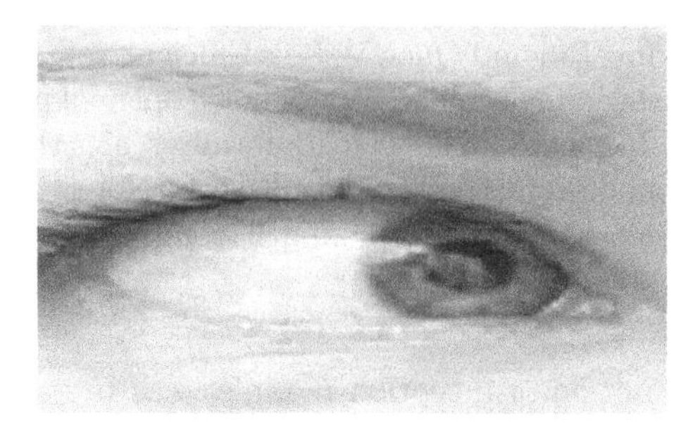

Fig. 3. Open Eye

Fig. 4. No Yawn

Fig. 5. Yawn

FER2013 Dataset. The FER2013 (Facial Expression Recognition 2013) dataset includes pictures and categories that indicate the subject's emotions. [3] The 48 by 48 pixel grayscale images in the collection depict seven distinct emotions: angry, disgusted, afraid, happy, sad, surprised, and neutral. 28709 examples from the training set, 3589 examples from the public testing set, and 3589 examples from the private test set make up the dataset.

Drivers' Survey Data. For conducting this survey, both oral interviews and google form were used. We asked drivers about their opinions on the reason of accident. We collected data to make our system more fit for users. There were 25 participants from different educational backgrounds and ages. We asked them different questions like what affects them most during driving, according to their observation what are the main reasons of accidents and whether a continuous monitoring system will help them or not.

3.3 Real World Implementation

The drowsiness and emotion of drivers will be detected in real world using raspberry pi. Python environment and necessary libraries like OpenCV, TensorFlow, Keras are installed in a raspberry pi for this purpose. We will have to integrate the drowsiness detection and emotion detection algorithms into a single Python script or application to run the code from raspberry pi. Using the camera of the raspberry pi, images will be captured from the camera module at regular intervals. Each image will be processed using the implemented algorithms to detect drowsiness and emotion. There will be a feedback mechanism to alert the driver in case of detected drowsiness or extreme emotions. We will use an alert system which will generate loud sound if it detects drowsiness or extreme emotion. The system will be tested under different lighting conditions, driver positions, and scenarios. At last, the system will be deployed in a vehicle or a simulated driving environment.

4 Result

4.1 Yawn_eye_dataset_new Dataset

Basic CNN, ResNet50, VGG16, VGG19 and InceptionV3 are applied in the dataset to make a comparison between their accuracy. Loss Graph and Accuracy Graph for each model are shown below:

Basic CNN: The loss graph and accuracy graph for Basic CNN model is shown by Fig. 6 and Fig. 7

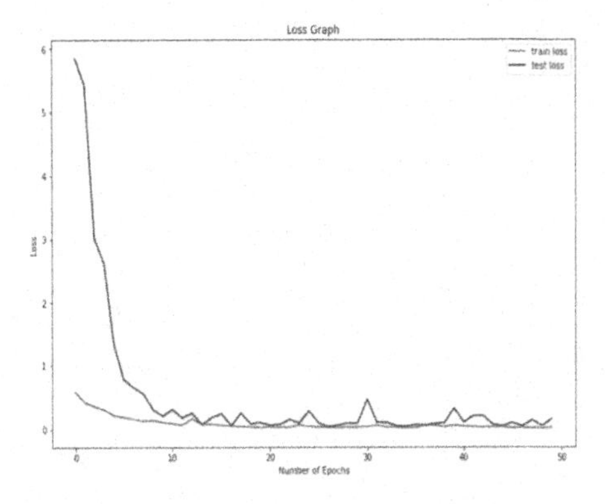

Fig. 6. Loss Graph of Basic CNN Model

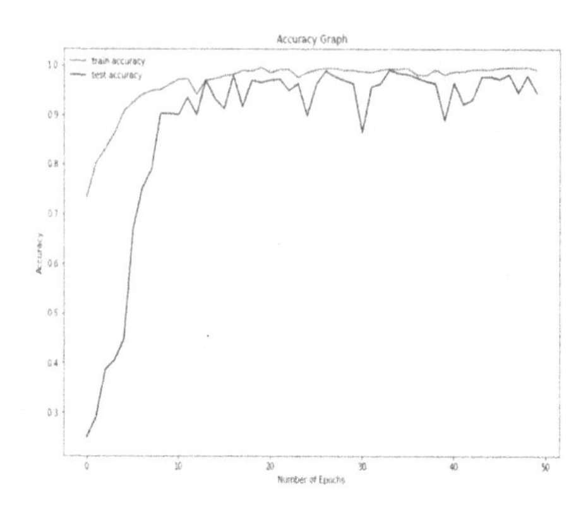

Fig. 7. Accuracy Graph of Basic CNN Model

ResNet50: The loss graph and accuracy graph for ResNet50 model is shown by Fig. 8 and Fig. 9

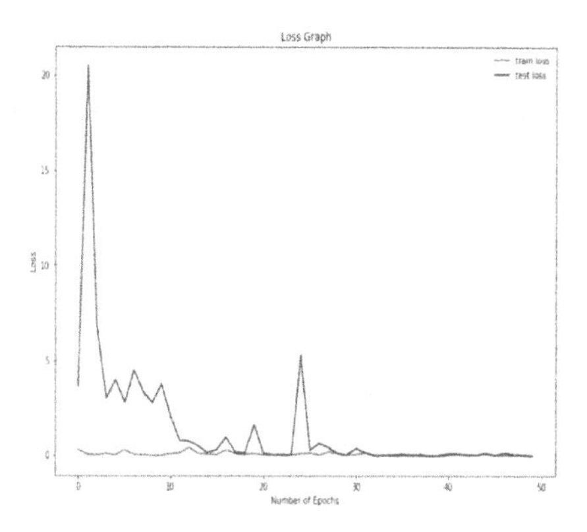

Fig. 8. Loss Graph of ResNet50 Model

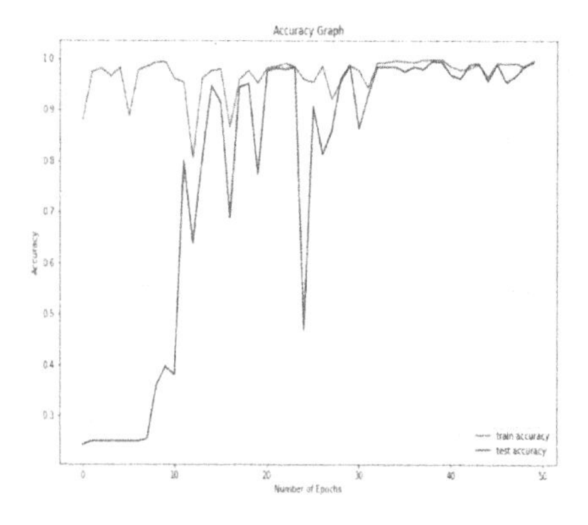

Fig. 9. Accuracy Graph of ResNet50 Model

VGG16: The loss graph and accuracy graph for VGG16 model is shown by Fig. 10 and Fig. 11

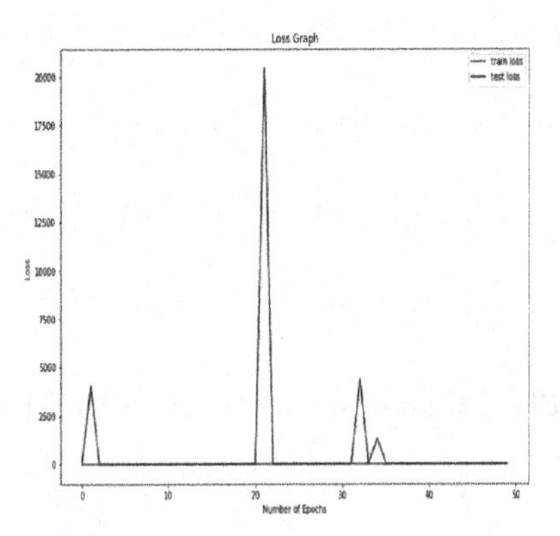

Fig. 10. Loss Graph of VGG16 Model

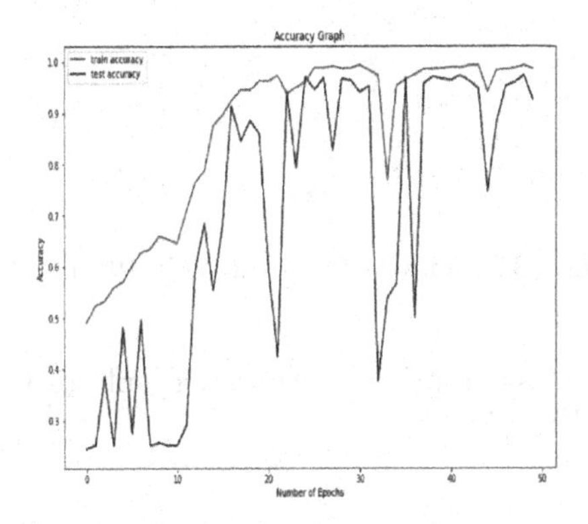

Fig. 11. Accuracy Graph of VGG16 Model

VGG19: The loss graph and accuracy graph for VGG19 model is shown by Fig. 12 and Fig. 13

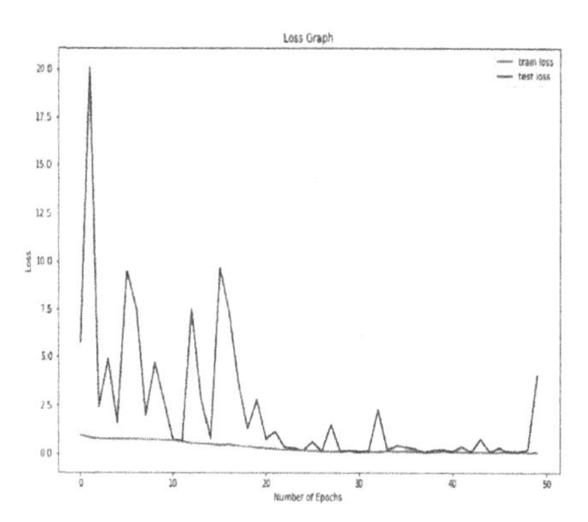

Fig. 12. Loss Graph of VGG19 Model

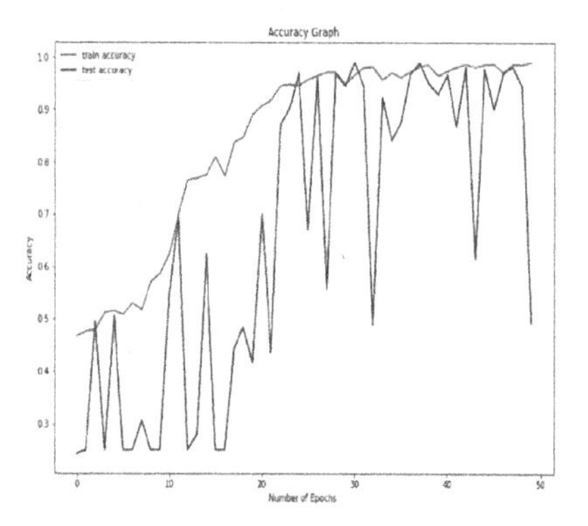

Fig. 13. Accuracy Graph of VGG19 Model

InceptionV3: The loss graph and accuracy graph for InceptionV3 model is shown by Fig. 14 and Fig. 15

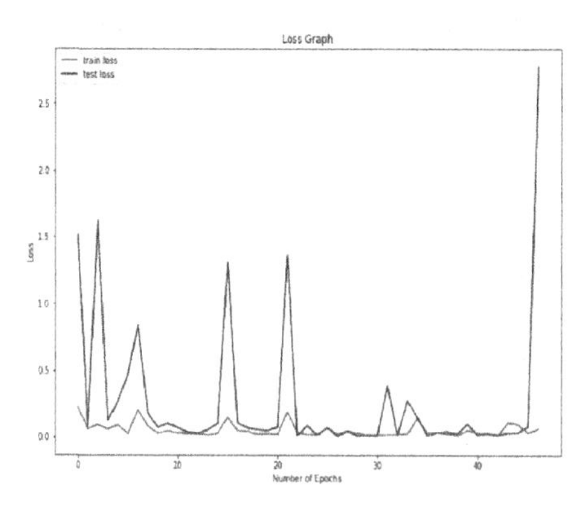

Fig. 14. Loss Graph of InceptionV3 Model

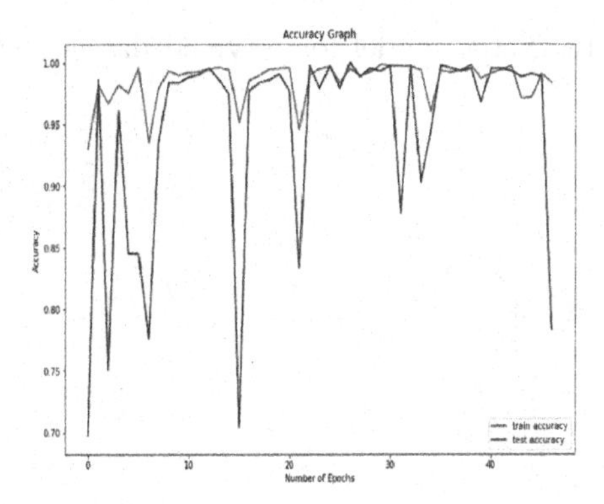

Fig. 15. Accuracy Graph of InceptionV3 Model

From the five model, the best accuracy is found from InceptionV3. It gives 100% accuracy. The second best accuracy is found from ResNet50 model. The accuracy table of the yawn_eye_dataset_new Dataset for different CNN models is shown by Table 1.

Table 1. Accuracy of different models of CNN

Model Name	Accuracy (%)
Basic CNN	94.46%
ResNet50	99.3%
VGG16	97.229%
VGG19	99.076%
InceptionV3	100%

4.2 Nitymed Dataset

ResNet50 is applied in this dataset which resulted in 100% accuracy. The precision, recall, and F-score achieved by the model are all 100%. Loss Graph and

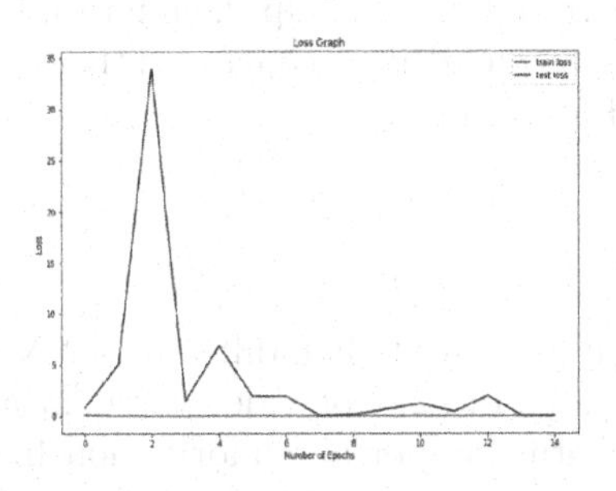

Fig. 16. Loss Graph of Nitymed Dataset for ResNet50 Model

Accuracy Graph for ResNet50 model for nitymed dataset are shown below by Fig. 16 and Fig. 17

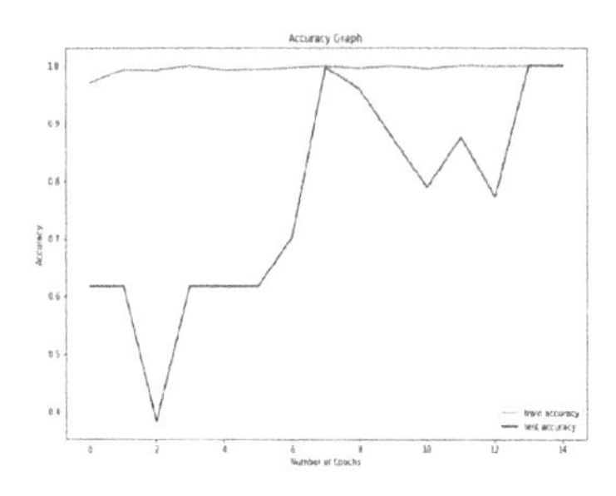

Fig. 17. Accuracy Graph of Nitymed Dataset for ResNet50 Model

4.3 FER2013 Dataset

In case of emotion detection, we experimented with FER2013 Dataset. The accuracy of this dataset is compared to low with other datasets. 64.80% accuracy was gained using Basic CNN, 62% accuracy was gained by ResNet50. The highest accuracy was found using InceptionV3 model which was 68.20%. In our future work we will try to improve the accuracy level.

4.4 Drivers' Survey Data

We conducted the survey on 25 drivers who drive different types of vehicles in Bangladesh. 87% of them were male and rest are female. Our question towards them was what affects them most at the time of driving. About 60% answered that they are affected by traffic jam, 33% indicated lack of sleep as another reason and the rest voted for several other reasons. According to 67% drivers the main reason of accidents are the breakings of traffic rules and overtaking tendencies. 40% indicated restless driving as another reason. There were options for selecting multiple reasons for drivers for accidents in the form. About 35% also indicated consumption of alcohol as a reason for accident. About 90% of drivers agreed that an alert system will help them for safe driving and it will also be helpful drowsy drivers. 80% of drivers concluded that a continuous monitoring system will be beneficial for them.

5 Discussion

From our experiments, it is proved that different CNN model works well for both image and video data. In case of video data, meaningful features from the video frames that capture important information have been extracted for classification. We also worked with detection of emotion. The goal was to create an detection method to use with raspberry pi. We wanted to work in agile method. By collecting user experiences we would like to make our system more useful. That's why we conducted a survey. Some important aspects came out from this survey like almost 90% drivers believes that a continuous monitoring system is necessary for road safety which gives our work validation. Another important thing was that we asked them what should be the range of cost to have an monitoring system for their vehicle. We designed the monitoring system using raspberry pi. The total system will cost approximately 50–75 dollar. But during our survey, we found that the drivers will like to have a system in the price range of 10–20 dollar. As we are working in an iterative approach, we will try to reduce the cost to meet the user's budget. In our future work we might have to use another tool instead of raspberry pi to reduce the cost.

6 Conclusion

The goal of this research is to put in place a system that will lessen traffic accidents. Various tests were carried out to detect emotions in drivers and their level of sleepiness. A study has been conducted to determine the drivers' preferences for monitoring systems. If the price falls within their budget, it appears that the majority of them would like to have one. In our next research, we hope to increase the accuracy of emotion detection and make a more affordable system proposal than the one we have now.

References

1. Raju, S.: Yawn-Eye Dataset New (2020). https://www.kaggle.com/datasets/serenaraju/yawn-eye-dataset-new. Accessed on [Insert Access Date]
2. Petrellis, N., Voros, N., Antonopoulos, C., Keramidas, G., Christakos, P., Mousouliotis, P.: Nitymed (2022). https://doi.org/10.34740/KAGGLE/DSV/3921886. https://www.kaggle.com/dsv/3921886
3. Goodfellow, I.J., et al.: FEr2013 (2013). https://www.kaggle.com/c/challenges-in-representation-learning-facial-expression-recognition-challenge/data. Accessed on [Insert Access Date]
4. Chand, H.V., Karthikeyan, J.: Cnn based driver drowsiness detection system using emotion analysis. Intell. Autom. Soft Comput. **31**(2) (2022)
5. El-Nabi, S.A., El-Shafai, W., El-Rabaie, E.-S.M., Ramadan, K.F., Abd El-Samie, F.E., Mohsen, S.: Machine learning and deep learning techniques for driver fatigue and drowsiness detection: a review. Multimed. Tools Appl., 1–37 (2023)
6. Fouad, I.A.: A robust and efficient eeg-based drowsiness detection system using different machine learning algorithms. Ain Shams Eng. J. **14**(3), 101895 (2023)
7. Florez, R., Palomino-Quispe, F., Coaquira-Castillo, R.J., Herrera-Levano, J.C., Paixão, T., Alvarez, A.B.: A cnn-based approach for driver drowsiness detection by real-time eye state identification. Appl. Sci. **13**(13), 7849 (2023)
8. Ahmed, M.I.B., et al.: A deep-learning approach to driver drowsiness detection. Safety **9**(3), 65 (2023)

9. Park, S., Pan, F., Kang, S., Yoo, C.D.: Driver drowsiness detection system based on feature representation learning using various deep networks. In: Asian Conference on Computer Vision, pp. 154–164. Springer (2016)
10. Chirra, V.R.R., Uyyala, S.R., Kolli, V.K.K.: Deep CNN: a machine learning approach for driver drowsiness detection based on eye state. Rev. d'Intelligence Artif. **33**(6), 461–466 (2019)
11. Zhao, Z., Zhou, N., Zhang, L., Yan, H., Xu, Y., Zhang, Z., et al.: Driver fatigue detection based on convolutional neural networks using EM-CNN. Computational intelligence and neuroscience 2020 (2020)

Improving Time to Take Over Through HMI Strategies Nudging a Safe Driving State

Roberta Presta(✉), Chiara Tancredi, Flavia De Simone, Mirko Iacono, and Laura Mancuso

Suor Orsola Benincasa University, Naples, Italy
{roberta.presta,chiara.tancredi}@unisob.na.it,
flavia.desimone@docenti.unisob.na.it,
{mirko.iacono,laura.mancuso}@studenti.unisob.na.it

Abstract. In scenarios of partially autonomous driving, drivers can easily become distracted and engage in secondary activities unrelated to driving. However, when it becomes necessary to take-over the control, they must be in the right conditions to resume driving safely. This article explores the design and evaluation of a Human-Machine Interface (HMI) that, leveraging the knowledge of the driver's state enabled by intelligent driver monitoring systems, helps the driver stay focused on the road in case of distraction and relax in case of agitation. The article presents the results of an experimental campaign in a driving simulator with 11 participants, aimed at measuring the advantage in terms of take-over reaction times with or without the proposed solution. This advantage was measured in terms of hands-on-wheel time through video analysis and eyes-on-road time through eye-tracking data analysis. The results demonstrate significant advantages both in terms of reaction times and from the perspective of user experience. The study allowed the identification of new research insights for further exploration of interface strategy.

Keywords: Driver Monitoring System · Human Machine Interface · User test · Take-Over Request · Eye-Tracking

1 Introduction

Driving safety remains a prevalent concern due to the persistently high number of road accidents. Furthermore, over 90% of these accidents result from human errors, often stemming from altered emotional states and distractions [1,6,17, 26,38].

This study is part of the NextPerception project that has received funding from the European Union Horizon 2020, *ECSEL* – 2019 – 2 – *RIA* Joint Undertaking (Grant Agreement Number 876487).

H. Krömker (Ed.): HCII 2024, LNCS 14732, pp. 27–43, 2024.
https://doi.org/10.1007/978-3-031-60477-5_3

For these reasons and owing to recent technological advancements, driving safety systems play a crucial role. Today, Advanced Driver Assistance Systems (ADAS) have the capability to assume control over both longitudinal (e.g., Adaptive Cruise Control) and lateral (e.g., Lane Keeping Assistant) guidance, or even both (e.g., Traffic Jam Assistant) [15]. This level of automation falls under SAE partial automation level (level 2), which mandates the driver to perform part of the dynamic driving task [34]. The next level of automation (SAE level 3) is conditional automation, where the automated driving system can perform the entire dynamic driving task within specific operational perimeters (e.g., on the highway). When a conditionally automated driving is engaged, the driver is not obligated to continuously supervise the system and can engage in non-driving related task (NDRT) [34]. However, when automation encounters its operational limits (e.g., exiting the highway or facing an unfamiliar situation), it alerts the driver to resume control by means of the activation of a take-over request (TOR). During the take-over process, the driver must swiftly shift attention to the driving scenario, regain physical control of the steering wheel and assess the situation to respond appropriately. It is essential that the take-over occurs quickly and effectively: the driver should not only be ready to take the control back, but should also be in the best possible state from both an attention and emotional standpoint to best regain control of driving [7,12,18,27,30,37,38].

Driver Monitoring Systems (DMSs), which collect extensive data on driver state and behavior can help in the assessment of the driver state, thus allowing to exploit such an information to design safer in-car interactions. Leveraging advanced sensors and artificial intelligence, by gathering data on drivers' facial expressions, body temperature, vocal tone, heart rate, and movement patterns, state-of-the-art DMSs can distinguish different aspects of the driver's state related to attention and emotions [2,9,28]. Developing human-machine interface (HMI) strategies that leverage the identification of unsafe states and subsequently support the driver in safe driving represents an important opportunity to be taken not only in L2 driving [24,30], but also in conditionally automated driving to prepare drivers at best for the take over process.

In this work, we present the design and the assessment by means of an experimental campaign of DMS-enabled interaction strategies in this context developed within the research effort of a recently concluded European project (NextPerception[1]). Specifically, the HMI designed in the NextPerception project is aimed at nudging the driver to maintain a safe driving state, that is, to pay attention to the road and counteract states characterized by too high a level of activation, as these have been shown to be suboptimal for safe driving. By means of the experimental study in a driving simulator environment presented in this paper, we want to show how much the use of this interface, based on the information provided by a DMS, can improve the recovery time of vehicle control in partially autonomous driving (L3) scenarios. Two types of measurements were considered to measure the drivers reaction times (RT) to the take-over request: (i) RT Hands-on stearing wheel, extracted from the analysis of the test video record-

[1] https://www.nextperception.eu/.

ings framing the steering wheel, and (ii) RT Eyes-on-Road, measured leveraging a wearable eye-tracking worn by the participants during the test. The work investigates also the results obtained in terms of user experience of the proposed support system leveraging self-judgments of the test participants.

The remaining sections of the article are organized as follows. The next section provides an overview of DMS-enabled HMI strategies implemented within the European *NextPerception* project to provide the reader with the research context and with details about the design of the *NextPerception* Take-Over Request HMI. Section 3 details the methodology applied for time measurement and TOR evaluation in the two comparing driving conditions (with NextPerception HMI strategy and without). Finally, Sect. 4 and Sect. 5 present the results and the discussion of the experimental assessment, and Sect. 6 offers final observations and design directions.

2 The Next Perception DMS-Enabled HMI Strategies and the TOR Design

Driver Monitoring Systems (DMS) aim to detect psychophysiological aspects of the driver early on to ensure safety during both manual and partially autonomous driving [19,21,31]. This information is utilized to adapt the vehicle's human-machine interface (HMI), enhancing communication with the driver. For instance, it can redirect visual attention to the road when the driver is distracted or adjust negative emotional conditions impacting driving performance [6,16,38]. Visual and cognitive distraction, along with emotional states characterized by extremely high arousal, such as anger, anxiety, euphoria, or extremely low ones often associated with drowsiness, negatively influence driver behavior and judgment in driving situations [12,26,38].

As remarked in [7,30], the optimal driving state is identified in the so-called "safe zone", characterized by average arousal and positive valence for the driver's emotional components [5,7,33], along with distraction avoidance [32].

Notably, the *NextPerception* Driver Monitoring System (DMS) identifies the Driver Complex State (DCS), encompassing cognitive and visual distraction, emotion type, and driver arousal, aiming to understand how best to assist them in safety-critical situations [9]. Based on this information, the system estimates the driver's ability to perform the driving task, communicated through an index ranging from 0 to 100, called Fitness to Drive [22]. If this index falls below 50%, the driver state is considered unfit. This data, combined with an estimate of the external driving environment, is leveraged by the Decision Support System (DSS). The DSS determines the most appropriate action to support the user, ranging from initiating an HMI recovery strategy to calm down or refocus the driver on road control, to hand over control of the vehicle from the driver to automation in the interest of safety [30].

2.1 *NextPerception* HMI Strategies

To nudge the driver in safe zone from the driving safety perspective, we designed three main HMI recovery strategies associated to different unsafe conditions that can be detected by the NextPerception DMS [30]:

- *Distraction (visual or cognitive)*: the strategy enabled by the DSS is refocusing.
- *High emotional arousal (e.g., in euphoric or angry states)*: the enabled strategy is relaxing.
- *Low emotional arousal (e.g., in pre-sleep states)*: the enabled strategy is activating.

Based on the detected combination of distraction with too-high or too-low arousal conditions, the refocusing strategies can be combined respectively with the relaxing and activating ones. Various channels have been exploited to implement the mentioned recovery strategies of the *NextPerception* HMI (Fig. 1) [30,32]:

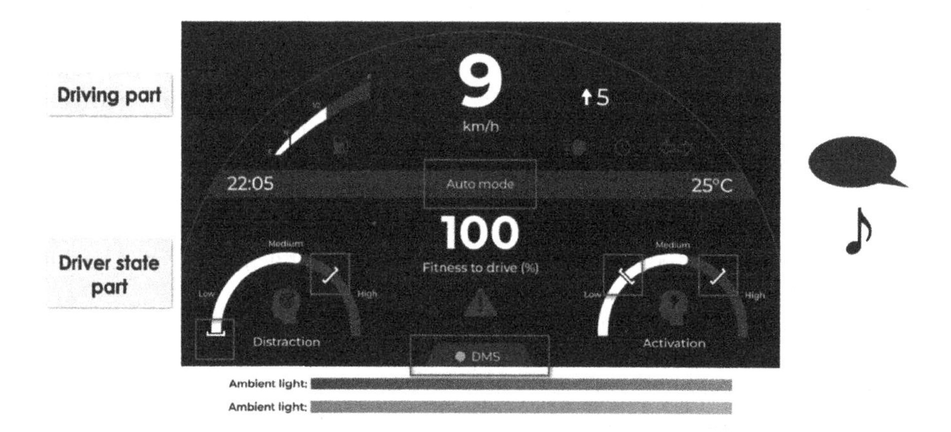

Fig. 1. Representation of the channels exploited to implement *NextPerception* HMI strategies: voice assistant (represented by the call-out in the picture), music (musical note), ambient light (respectively, blue for relaxing strategies and orange for activating strategies), and visual (displayed HMI). In the display, the bottom area is dedicated to show the level of distraction (on the left), the fitness to drive index (in the centre) and the level of activation (on the right). The square brackets in the distraction and activation indicators identify the driver's state "safe zone".

- *Ambient lights:* LEDs around the dashboard change color based on the driver's activation state. If the driver is too excited (high arousal or activation), the LEDs will light blue; if too unresponsive (low arousal or activation), the LEDs will light orange.

- *Music:* music can be effectively used to help drivers in calming down or to reactivate. According to user preferences, user pre-defined playlists (relaxing or energizing, respectively) can be proactively proposed by the HMI to the purpose.
- *Voice assistant:* for communicating empathetically with drivers and proactively supporting them in driving scenarios.
- *Visual HMI:* implemented in the dashboard, showing in the bottom part (from left to right in Fig. 1) the level of distraction, the fitness to drive index and the level of activation, besides the driving mode (auto or manual) and whether the DMS is activated or not [29]. The safe zone is represented by the squared brackets in the distraction and activation indicators: exceeding the safe zone is signaled by the indicator changing color and the icon turning orange. The fitness to drive indicator when entering an alarm state is also turning orange and causing the danger triangle to light up.

To show the HMI in action, we leverage one of the user scenarios developed in the *NextPerception* project, namely the one related to partially automated driving [28,30]. The protagonist, Julie, is driving on a highway in automatic mode. Julie's state is altered as she has been arguing on the phone with her boyfriend, resulting in high distraction, emotional activation, and negative emotional valence. This situation exemplifies an unsafe driver's state. The vehicle's automation, following the predefined destination, acknowledges that the moment to prompt Julie to regain control is approaching, as the highway exit is nearby. Thanks to the DMS (Driver Monitoring System), the automation also recognizes that Julie is unfit to drive, and that it is the case to apply a recovery strategy (refocusing + relaxing) using the HMI (Human-Machine Interface) to guide Julie back into the safe zone. Figure 2 shows how the HMI would appear:

- Blue ambient lights;
- Relaxing song;
- Voice assistant: *"Dangerous state detected, keep an eye on the road and relax"*;
- Visual HMI: activation indicator, distraction indicator, and fitness to drive indicator in an alarm state.

2.2 *NextPerception* Take-Over Request design

Previous studies [3,8,10–12,15,20,23,25,27,36,39] and the current implementation of the *NextPerception* HMI were taken into consideration for the design of the Take-Over Request to maintain necessary consistency. The TOR implementation was multimodal, as suggested by the study in [36], utilizing channels already employed for HMI strategies, namely:

- *Audio channel:* short alert sound.
- *Voice assistant:* used to communicate the request to resume control, employing the phrase *"Please resume control"* as suggested by [11].

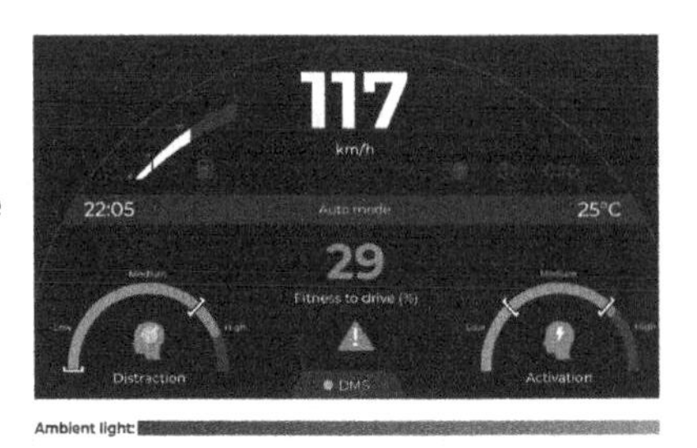

Fig. 2. Representation of Julie's user story and related HMI strategies (*relaxing+refocusing*), which also involves the use of relaxing music and voice assistant

– *Visual HMI:* In the center of the HMI, within the gray area that separates the driving and driver state sections, the time budget for the take-over is displayed. As illustrated in Fig. 1, the driving mode (Auto mode or Manual mode) was originally presented in this gray dividing area. When the Take-Over Request (TOR) is initiated, an icon depicting two hands on the steering wheel replaces the driving mode icon, synchronized with the alert beep. Subsequently, the term *"TOR"* appears beside the hands-on-the-steering-wheel icon, accompanied by a progressively decreasing linear bar indicating the available time budget for the take-over (10 s budget, as suggested by [23]). Upon reaching the end of this countdown, the transition to manual mode is activated (Fig. 3).

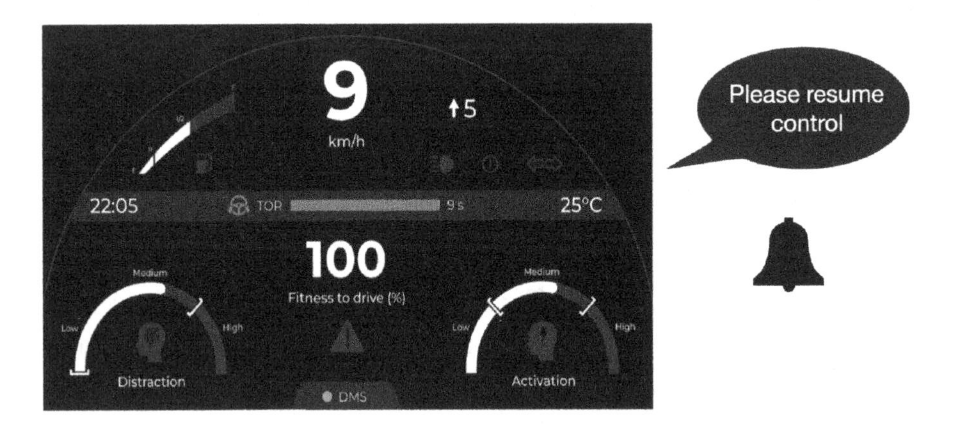

Fig. 3. *NextPerception* Take-Over Request strategy: visual HMI TOR with time budget, voice assistant and short alarm sound.

3 Methodology

The research question that moves the experimental campaign was the following: can the multimodal HMI nudging strategy, presented in the previous section, improve the take-over reaction times? To answer this question we designed a user test implemented at the driving simulator aimed at comparing the take-over reaction times with and without the NextPerception HMI. The experiment was a within study where each participant experienced both the HMIs in the same take-over scenarios. The two experimental conditions (driving with the NextPerception HMI: condition C, and driving without the NextPerception HMI: condition S) were tested on the same driving scenario on a medium-traffic highway in autonomous mode, divided into four one-minute phases (Fig. 5):

1. *Baseline*: observing autonomous driving without specific tasks;
2. *Non-driving related task (NDRT)*: participants played a video game on a smartphone (*Bunny Hammer*), inducing distraction and high activation because of the game time pressure
3. *Pre-take-over*: in this minute, participants involved in the S driving trial continued to perform the NDRT task as in the previous phase, without even realizing the phase shift. In contrast, participants involved in the C driving trial experienced the NextPerception HMI recovery strategy, nudging refocusing + relaxation: following the example depicted in Julie's scenario, the activation and distraction indicators go in the alarm mode, as well as the fitness to drive indicator; a blue LED light shows up and the voice assistant pronounces the message "Dangerous state detected, keep an eye on the road and relax".
4. *Take-over*: when entering this phase, the take-over request is presented to the participant, requiring participants to assume control of the vehicle and continue in manual driving until the end of the minute.

A recorded voice guided participants through the simulation phases. The participant sample was divided into two experimental groups (*group A:* first driving test: S, second driving test: C; *group B:* first driving test: C, second driving test: S), in order to balance order effects in the resulting take-over reaction times.

TOR timing was evaluated using two metrics:

- *Reaction times (RT) Hands-on:* the time (in seconds) between the TOR and the moment when the participant grips the steering wheel with both hands.
- *Reaction times (RT) Eyes-on-road:* the time (in seconds) taken to return gaze to the road for the first time (first fixation) following the TOR.

To collect this data, a camera was positioned to capture the participant and the steering wheel, enabling the measurement of RT Hands-on. Additional information was gathered through the *Tobii Pro Glasses 2* eye-tracking system, observing eye-gazing behavior during testing and calculating RT Eyes-on-Road (Fig. 4).

We were also interested in evaluating the user experience in the two different take-over modalities, and we utilized specific questionnaires for this purpose, as outlined in the procedure below.

Fig. 4. Driving simulator environment of the user test. To compute the RT Hands-on-wheel, the participant video was collected by means of the red smartphone on the top of the picture. Eye-tracker Tobii Pro Glasses 2, used to compute RT Eyes-on-road are also shown. (Color figure online)

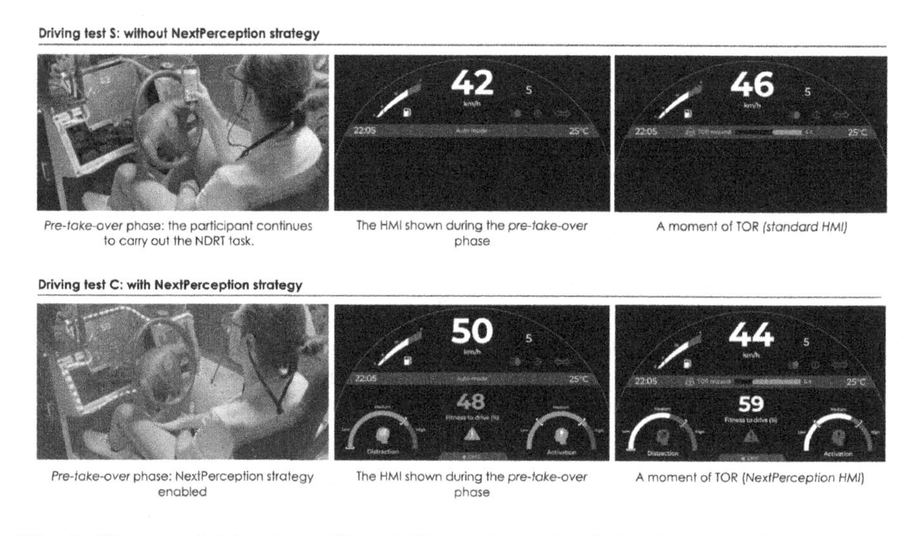

Fig. 5. The two driving tests (S and C) are shown in their phases. In driving test C, during the third phase (pre-take-over phase), the nudging HMI state-recovery strategy is activated and the behavior of the driver can change: they can freely decide to stop the NDRT task or not.

3.1 Procedure

An experimenter was responsible for guiding each participant through the experimental procedure, which was based on the following steps:

1. *Explanation of the NextPerception project and assignment of participant ID (5 min):* Upon arrival in the laboratory, participants were introduced to the NextPerception project and assigned an ID to anonymize the collected data.
2. *Privacy Information and Informed Consent (5 min):* Participants were explained that personal, physiological, driving, and feedback data would be collected anonymously and used only by the project researchers in aggregate form for research purposes. They were also informed that they could leave the test at any time and under any circumstances.
3. *Entry Questionnaire (10 min):* Participants were given a questionnaire to assess demographic data as well as knowledge and familiarity with the technology used in the experiment.
4. *Introduction to the driving test (5 min):* Participants were briefed on the various stages of the driving test. Regarding the third phase, participants were informed that, when testing the NextPerception HMI, the car could detect an altered state due to the distraction and activation induced by the NDRT (i.e., the videogame) and eventually activate specific interactions. They were also shown the videogame, and a playtest was conducted to let them familiarize with it.
5. *Free ride and eye-tracker calibration (5 min):* Once participants reached the simulator room, they performed a three-minute free ride to familiarize themselves with the simulator. The eye-tracking system, necessary for observing participants' visual behavior, was then calibrated by having participants wear it and carry out the appropriate procedure.
6. *Driving test (4 min):* This phase followed the driving phases explained above.
7. *Exit questionnaire (10 min):* An additional questionnaire was administered to participants to assess the user experience of the driving trial in each condition.

3.2 Entry Questionnaire

The entry questionnaire was designed to collect biographical data (age, education level, gender identity), years of driving experience, and experience and familiarity with technology in general (using the Affinity for Technology Interaction - ATI by [13]), and specifically with partially autonomous driving vehicles and DMS.

3.3 Exit Questionnaire

At the end of each trial, a concluding questionnaire was administered, asking participants for evaluations and information:

- To assess their user experience according to the UEQ-short form questionnaire [35]

- Their emotional state during various phases of the trial using the SAM questionnaire [4,14].
- The level of perceived readiness concerning take-over, through a scale from 1 to 10, adapted from the one used by [10].

3.4 Hypotheses and Data Analysis

The presented study aimed to validate the following hypotheses:

H1: Through the use of the refocusing and relaxing strategies designed for *NextPerception*, which function as a nudge, "gently pushing" the driver back into a safer driving state [30,32] before the take-over, the driver is induced to regain control more quickly than when such a strategy is not applied. To test this hypothesis, three hypotheses were considered in the analysis of specific data:
H1a: RT Hands-on [C] < RT Hands-on [S]
H1b: RT Eyes-on-road [C] < RT Eyes-on-road [S]

H2: The Next Perception HMI enables the driver to be more prepared for the take-over compared to the alternative scenario. To test this hypothesis, we compared the results of self-assessments obtained through an ex-post questionnaire to verify whether TOR readiness [C] < TOR readiness [S]..

H3: The proposed NDRT induced the expected emotion (high activation and positive valence), and thus, although the detection of the emotional state was simulated, the driving test was perceived as realistic. We can formulate this hypothesis as follows: SAM baseline [C+S] < SAM NDRT [C+S]

H4: The *refocusing+relaxing* strategy succeeds in making the driver concentrate and calm down. We can formulate this hypothesis as follows:
H4a: the majority of participants spontaneously interrupted the execution of the NDRT following the activation of the nudging HMI
H4b: SAM NDRT [C] > SAM pre-take-over [C], meaning that the intensity of the arousal and of the positive valence is lower after the application of the HMI

H5: The user experience of driving with the NextPerception strategy [C] is rated more positively than without it: UEQ-s [C] > UEQ-s [S].

Regarding the data collection of RT Hands-on and RT Eyes-on Road, video analysis and time measurement were carried out using *Noldus' The Observer XT v.17* software, while *IBM® SPSS® Statistics* was used for the statistical analysis.

4 Results

Before embarking on data collection, we measured the sample size needed to have reliable results for the take-over reaction times, which was 8 participants. The study was conducted in July 2023, and 13 people initially participated. However, two participants were excluded due to motion sickness issues. This brought the total number of participants to 11 (5 women, 6 men), aged between 25 and 40 years (M = 28.27; SD = 4.10). Among them, 4 had 10 years or more of driving experience, 6 had 3–10 years of driving experience, and 1 had less than 3 years of driving experience. In general, participants exhibited average levels of openness to technology use (ATI results: M = 3.77, range 1–6). Regarding familiarity and experience with driving monitoring systems, participants reported average values (mean familiarity = 5.27; mean experience = 2.09; range 1–7). Similar values were recorded for autonomous driving (mean familiarity = 5.82; mean experience = 1.91; range 1–7).

The participants were evenly divided into two experimental groups, with Group A consisting of 5 participants and Group B consisting of 6 participants. Figure 6 shows a participant during both drives.

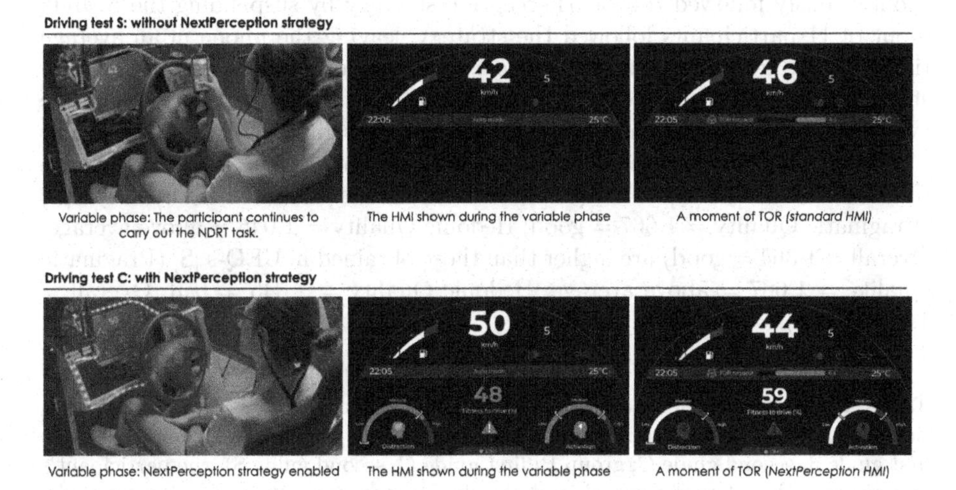

Fig. 6. The third minute of the driving tests (*Pre-take-over phase*) is shown: in guide S the participant continues to play because there are no HMI changes with respect to the previous phase, while in guide C the NextPerception strategy is enabled. The respective visual HMIs during the pre-take-over phase and take-over phase are shown.

Regarding the TOR results, we can confirm the hypothesis that the NextPerception strategy allows for shorter take-over processes (**H1**). This is evident as the TOR times in the C driving conditions are significantly shorter than those in the S driving condition. Specifically, hypothesis **H1a:** RT Hands-on [C] < RT Hands-on [S] is confirmed. The difference is statistically significant in reaction times Hands-on (C: M = 1.49, S: M = 4.65, $t(10) = 4.19$, $p < 0.0019$). Similarly, the reaction times Eyes-on also exhibit a statistically significant difference (C: M = 0.43, S: M = 2.66, $t(10) = 3.43$, $p < 0.0064$), confirming **H1b:** RT Eyes-on-road [C] < RT Eyes-on-road [S]. Table 1 shows the RT Hands-on and RT Eyes-on-road divided by guides (S and C) for each participant.

Regarding **H2:** TOR readiness [C] > TOR readiness [S], the difference is not statistically significant (C: M = 7.3, S: M = 6.9, range 1–10), as participants shown a good TOR readiness in both cases.

H3: The simulated driving scenario in C-driving trials seemed credible to participants, as supported by Hypothesis 3. Indeed, H3 appears to be verified. SAM baseline [C+S] > SAM NDRT [C + S] is confirmed in all three dimensions (valence, arousal, dominance) of the SAM ($F(2,40) = 4.20$ $p < 0.022174$ SS = 2.92 MSe = 0.35). That means that participants experienced the desired state conditions while in the NDRT phase, and that the HMI intervention for C driving trials was then appropriated even if simulated.

H4: the refocusing + relaxing HMI shown positive effects for the refocusing strategy and non significant effect for the relaxing part. Indeed, we conducted an observational study of participants while driving C to see how many had spontaneously followed the NextPerception strategy by suspending the NDRT: 9 out of 11 participants followed the strategy, leaving the phone in an average time of 9.59 s from the start of the vocal assistant nudge of the NextPerception strategy and focusing on the road (**H4a**). **H4b** (SAM NDRT [C] > SAM pre-take over phase [C]), on the other hand, did not yield a significant result.

Finally, analyzing the UEQ-s data using the tools provided by [35], we can confirm **H5:** UEQ-s [C] > UEQ-s [S]. Indeed, the results obtained in UEQ-s [C] (Pragmatic Quality = 1.667 = good; Hedonic Quality = 1.045 = above average; Overall = 1.364 = good) are higher than those obtained in UEQ-s [S] (Pragmatic Quality = 1.667 = above average; Hedonic Quality = 1.045 = bad; Overall = 1.364 = bad).

Table 1. Reaction Times Hands-on and Reaction Times Eyes-on-road for each participant are shown. The participant ID is formed by the participant's group (group A: first guide S, second guide C; group B: first guide C, second guide S), the participant's number in order of performance, and the driving test type (S or C). For example, B2C is the driving test with the NextPerception strategy carried out by the second participant in group B.

Drive tests [C]	RT Hands-on (seconds)	RT Eyes-on (seconds)	Drive tests [S]	RT Hands-on (seconds)	RT Eyes-on (seconds)
A1C	1.57	1.12	A1S	3.57	3.12
A3C	3.37	2.00	A3S	3.87	2.24
A4C	0.00	0.00	A4S	3.30	2.40
A5C	0.00	0.48	A5S	3.54	0.04
A6C	0.00	0.00	A6S	5.71	4.96
A7C	2.67	0.00	A7S	6.17	2.04
B1C	1.23	0.00	B1S	2.37	315
B2C	4.57	0.84	B2S	4.10	056
B4C	2.94	0.00	B4S	5.11	1.64
B5C	0.00	0.28	B5S	5.21	2.36
B6C	0.00	0.00	B6S	8.17	6.72
Mean	1.49	0.43	**Mean**	4.65	2.66
SD	1.66	0.65	**SD**	1.63	1.88

5 Discussion

The focus of the study was primarily on investigating the impact of the NextPerception strategy on Take-Over Request (TOR) times and driver behavior. The results confirmed the hypothesis that the NextPerception HMI strategy significantly improves take over reaction times when compared with a standard HMI strategy. TOR readiness self-assessments by participants reflected high average values for both conditions (C: M = 7.3, S: M = 6.9), suggesting that participants did not feel particular difficulties in taking over the vehicle control in both conditions.

The absence of a significant difference between the two measurements could be attributed not only to the sample size but also to the driving scenario not encompassing challenging conditions when the Take-Over Request (TOR) is issued, such as an emergency maneuver due to an accident or a detour for construction. Consequently, this scenario may not induce feelings of urgency or danger for the drivers.

Regarding the induction of emotional states, with the confirmation of **H3**: SAM baseline [C + S] > SAM NDRT [C+S], we can assert that the Non-Driving Related Task (NDRT) altered participants' state in the desired way, i.e., increasing both the valence and the arousal of their emotional state, providing alongside distraction. Consequently, we can conclude that the activation of the NextPerception strategy was perceived as realistic and driven by participants' emotional and distracted states, despite being simulated. However, we cannot affirm that the relaxing strategy changed the emotional state of participants; rather, it succeeded in refocusing drivers on the driving task. This is further supported by the fact that 9 out of 11 participants put down their phones and focused on the road.

Finally, participants expressed a more positive experience with the NextPerception strategy, as reflected in higher UEQ-s scores for Pragmatic Quality, Hedonic Quality, and Overall Experience.

6 Conclusion and Future Work

This research work documents the design and evaluation of an HMI (Human-Machine Interface) strategy conceived to encourage drivers to be in a safe driving state before taking control of the vehicle in partially automated driving (NextPerception HMI). The HMI strategy leverages the driver information state provided by DMSs (Driver Monitoring Systems) capable of detecting driver distraction and activation, such as the DMS realized in the European NextPerception project. Specifically, the proposed HMI is designed to be applied when drivers are distracted and highly activated by NDRT (Non-Driving Related Task), and the time to perform the take-over request is approaching. A multimodal combination of visual alarms, ambient lighting, music, and a vocal assistant has been prototyped to nudge the driver to refocus on the road and relax.

To examine the effects of the HMI nudging strategy on take-over reaction times, we conducted an experimental campaign in a driving simulator environment in a partially automated driving trial. Participants were involved in a distracting and highly activating NDRT when the take over request was issued. They experienced two different driving trials in the same scenario, both with the NextPerception HMI and without. Using video analysis and eye-tracking data, we computed the Hands-on-Wheel reaction times and Eyes-on-Road reaction times for 11 participants in two driving trial conditions: with the NextPerception HMI and without the NextPerception HMI. We also took the opportunity to investigate the perceived user experience and the perceived time to take-over readiness in both cases.

Statistical analysis confirmed that when the nudging HMI strategy is applied (aimed at refocusing and relaxing the driver), take-over times are significantly reduced, and it is effective in recalling the user's attention to the road in 9 out of 11 cases. While the refocusing effect seems effective and promising, the relaxing effect we measured was not statistically significant and needs further exploration, considering the enlargement of the participant sample. Nonetheless, the user experience in the take-over process was rated better when the NextPerception HMI was involved than in the other case.

In conclusion, participants show positive attitudes towards being nudged to adopt safer behavior even during autonomous driving, and the improvement in terms of take-over reaction time is significant. This paves the way for further investigation in more ecological contexts and more articulated driving scenarios to determine the timing and circumstances when such interaction strategies should be triggered. The modalities of the relaxing strategy would also benefit from feedback coming from a bigger participant sample. Separate test should be conducted to validate the relaxing (and dually also the activating) effects of the envisioned nudging strategy when partially automated driving is involved.

References

1. NHTSA - National Highway Traffic Safety Administration: 2016 fatal motor vehicle crashes: overview. Traffic safety facts: research note. Report No. DOT HS 812 456 (2017)
2. Amparore, E.G., Botta, M., Drago, I., Donatelli, S., Mazzone, G.: Challenges for driver action recognition with face masks. In: 2022 IEEE 25th International Conference on Intelligent Transportation Systems (ITSC), pp. 1491–1497. IEEE (2022)
3. Borojeni, S.S., Chuang, L., Heuten, W., Boll, S.: Assisting drivers with ambient take-over requests in highly automated driving. In: Proceedings of the 8th International Conference on Automotive User Interfaces and Interactive Vehicular Applications, pp. 237–244 (2016)
4. Bradley, M.M., Lang, P.J.: Measuring emotion: the self-assessment manikin and the semantic differential. J. Behav. Ther. Exp. Psychiatry **25**(1), 49–59 (1994)
5. Braun, M., Schubert, J., Pfleging, B., Alt, F.: Improving driver emotions with affective strategies. Multimodal Technol. Interact. **3**(1), 21 (2019)

6. Braun, M., Weber, F., Alt, F.: Affective automotive user interfaces-reviewing the state of driver affect research and emotion regulation in the car. ACM Comput. Surv. (CSUR) **54**(7), 1–26 (2021)
7. Cai, H., Lin, Y.: Modeling of operators' emotion and task performance in a virtual driving environment. Int. J. Hum. Comput. Stud. **69**(9), 571–586 (2011)
8. Choi, D., Sato, T., Ando, T., Abe, T., Akamatsu, M., Kitazaki, S.: Effects of cognitive and visual loads on driving performance after take-over request (TOR) in automated driving. Appl. Ergon. **85**, 103074 (2020)
9. Davoli, L., et al.: On driver behavior recognition for increased safety: a roadmap. Safety **6**(4), 55 (2020)
10. Du, N., et al.: Evaluating effects of cognitive load, takeover request lead time, and traffic density on drivers' takeover performance in conditionally automated driving. In: 12th International Conference on Automotive User Interfaces and Interactive Vehicular Applications, pp. 66–73 (2020)
11. Eriksson, A., Stanton, N.A.: Takeover time in highly automated vehicles: noncritical transitions to and from manual control. Hum. Factors **59**(4), 689–705 (2017)
12. Forster, Y., Naujoks, F., Neukum, A., Huestegge, L.: Driver compliance to take-over requests with different auditory outputs in conditional automation. Accid. Anal. Prev. **109**, 18–28 (2017)
13. Franke, T., Attig, C., Wessel, D.: A personal resource for technology interaction: development and validation of the affinity for technology interaction (ATI) scale. Int. J. Hum.-Comput. Interact. **35**(6), 456–467 (2019)
14. Geethanjali, B., Adalarasu, K., Hemapraba, A., Pravin Kumar, S., Rajasekeran, R.: Emotion analysis using SAM (self-assessment manikin) scale. Biomed. Res. (0970-938X) **28** (2017)
15. Gold, C., Damböck, D., Lorenz, L., Bengler, K.: "take over!" How long does it take to get the driver back into the loop? In: Proceedings of the Human Factors and Ergonomics Society Annual Meeting, vol. 57, pp. 1938–1942. Sage Publications Sage CA, Los Angeles, CA (2013)
16. Horberry, T., et al.: Human-centered design for an in-vehicle truck driver fatigue and distraction warning system. IEEE Trans. Intell. Transp. Syst. **23**(6), 5350–5359 (2021)
17. Jeon, M.: Emotions and affect in human factors and human–computer interaction: taxonomy, theories, approaches, and methods. In: Emotions and Affect in Human Factors and Human-Computer Interaction, pp. 3–26 (2017)
18. Kerschbaum, P., Lorenz, L., Hergeth, S., Bengler, K.: Designing the human-machine interface for highly automated cars-challenges, exemplary concepts and studies. In: 2015 IEEE International Workshop on Advanced Robotics and Its Social Impacts (ARSO), pp. 1–6. IEEE (2015)
19. Koesdwiady, A., Soua, R., Karray, F., Kamel, M.S.: Recent trends in driver safety monitoring systems: state of the art and challenges. IEEE Trans. Veh. Technol. **66**(6), 4550–4563 (2016)
20. Louw, T.L., Merat, N., Jamson, A.H.: Engaging with highly automated driving: to be or not to be in the loop? In: 8th International Driving Symposium on Human Factors in Driver Assessment, Training and Vehicle Design, Leeds (2015)
21. Manstetten, D., et al.: The evolution of driver monitoring systems: a shortened story on past, current and future approaches how cars acquire knowledge about the driver's state. In: 22nd International Conference on Human-Computer Interaction with Mobile Devices and Services, pp. 1–6 (2020)

22. Andruccioli, M., Mengozzi, M., Presta, R., Mirri, S., Girau, R.: Arousal effects on fitness-to-drive assessment: algorithms and experiments. In: 2023 IEEE 20th Annual Consumer Communications & Networking Conference (CCNC). IEEE (2023)
23. Melcher, V., Rauh, S., Diederichs, F., Widlroither, H., Bauer, W.: Take-over requests for automated driving. Procedia Manufact. **3**, 2867–2873 (2015)
24. Merat, N., Lee, J.D.: Preface to the special section on human factors and automation in vehicles: designing highly automated vehicles with the driver in mind. Hum. Factors **54**(5), 681–686 (2012)
25. Morales-Alvarez, W., Sipele, O., Léberon, R., Tadjine, H.H., Olaverri-Monreal, C.: Automated driving: a literature review of the take over request in conditional automation. Electronics **9**(12), 2087 (2020)
26. Neta, M., Cantelon, J., Haga, Z., Mahoney, C.R., Taylor, H.A., Davis, F.C.: The impact of uncertain threat on affective bias: individual differences in response to ambiguity. Emotion **17**(8), 1137 (2017)
27. Petermeijer, S., Doubek, F., De Winter, J.: Driver response times to auditory, visual, and tactile take-over requests: a simulator study with 101 participants. In: 2017 IEEE International Conference on Systems, Man, and Cybernetics (SMC), pp. 1505–1510. IEEE (2017)
28. Presta, R., Chiesa, S., Tancredi, C.: Driver monitoring systems to increase road safety. Hum. Body Interact., 247 (2022)
29. Presta, R., De Simone, F., Mancuso, L., Chiesa, S., Montanari, R.: Would I consent if it monitors me better? A technology acceptance comparison of BCI-based and unobtrusive driver monitoring systems. In: 2022 IEEE International Conference on Metrology for Extended Reality, Artificial Intelligence and Neural Engineering (MetroXRAINE), pp. 545–550. IEEE (2022)
30. Presta, R., De Simone, F., Tancredi, C., Chiesa, S.: Nudging the safe zone: design and assessment of HMI strategies based on intelligent driver state monitoring systems. In: Krömker, H. (ed.) HCII 2023. LNCS, vol. 14048, pp. 166–185. Springer, Cham (2023). https://doi.org/10.1007/978-3-031-35678-0_10
31. Presta, R., Tancredi, C., De Simone, F., Chiesa, S., Mancuso, L., Marino, L.: Training intelligent driver state monitoring systems: design and validation of an experimental procedure in a driving simulator environment. In: 2023 IEEE International Conference on Metrology for eXtended Reality, Artificial Intelligence and Neural Engineering (MetroXRAINE). IEEE (2023)
32. Presta, R., Tancredi, C., Mancuso, L.: In the loop of safe driving: an assessment of HMI strategies enabled by intelligent driver monitoring systems with daily drivers. In: 2023 IEEE International Conference on Metrology for eXtended Reality, Artificial Intelligence and Neural Engineering (MetroXRAINE). IEEE (2023)
33. Russell, J.A.: A circumplex model of affect. J. Pers. Soc. Psychol. **39**(6), 1161 (1980)
34. SAE International: Automated driving levels of drivings are defined in new SAE international standart j3016. Warrendale, AS (2014)
35. Schrepp, M., Hinderks, A., Thomaschewski, J.: Design and evaluation of a short version of the user experience questionnaire (UEQ-S). Int. J. Interact. Multimedia Artif. Intell. **4**(6), 103–108 (2017)
36. Wintersberger, P., Schartmüller, C., Sadeghian, S., Frison, A.K., Riener, A.: Evaluation of imminent take-over requests with real automation on a test track. Hum. Factors **65**(8), 1776–1792 (2023)

37. Zeeb, K., Buchner, A., Schrauf, M.: What determines the take-over time? An integrated model approach of driver take-over after automated driving. Accid. Anal. Prev. **78**, 212–221 (2015)
38. Zepf, S., Hernandez, J., Schmitt, A., Minker, W., Picard, R.W.: Driver emotion recognition for intelligent vehicles: a survey. ACM Comput. Surv. (CSUR) **53**(3), 1–30 (2020)
39. Zhang, B., De Winter, J., Varotto, S., Happee, R., Martens, M.: Determinants of take-over time from automated driving: a meta-analysis of 129 studies. Transport. Res. F: Traffic Psychol. Behav. **64**, 285–307 (2019)

Impacts of Training Methods and Experience Types on Drivers' Mental Models and Driving Performance

Linwei Qiao, Jiaqian Li, and Tingru Zhang(✉)

Institute of Human Factors and Ergonomics, College of Mechatronics and Control Engineering, Shenzhen University, Shenzhen, China
{2110292067,2310295039}@email.szu.edu.cn, zhangtr@szu.edu.cn

Abstract. Training for driver in automated vehicle (AV) systems is considered an effective way to establishing correct driver mental models and enhancing the safety of autonomous driving usage. With the increasing market share of Level 2 AV systems, it has become necessary to determine some efficient training methods of AV systems. For drivers, different experiences lead to different prior knowledge, which directly affects the formation of mental models. Therefore, this study explored the impact of training methods on the mental models and driving performance of drivers with different driving experience. We recruited 72 participants with three different types of driving experience. They were equally divided in one of the four training groups. After training, they would take a test drive in a driving simulator. Driving data and mental models were collected. The results indicated that both training methods and driving experience significantly affect the formation of mental models. Furthermore, for the scenarios within the training, the difference between training methods would be amplified with the increasing level of scenario urgency. And post-training driving experience would mitigate the differences between driving experience types. Video training enhanced drivers' attentiveness more than other three training methods. Lastly, for drivers who had not used AVs, having more extensive driving experience might reduce their attentiveness in hazardous road sections. These results highlight the differences in the effectiveness of training for drivers with different experience types and emphasize the need to develop training programs tailored to drivers' experience types to improve their mental models.

Keywords: mental model · prior knowledge · training · takeover quality

1 Introduction

With the advancement of AV systems, vehicles equipped with Level 2 AV systems have gained widespread use in people's daily lives. However, concurrently, accidents related to intelligent driving have become a topic of significant concern. Ganesh (2020) summarized multiple intelligent driving accidents, finding that a common cause of these incidents is the human drivers' attention diversion and their inability to take over control

H. Krömker (Ed.): HCII 2024, LNCS 14732, pp. 44–56, 2024.
https://doi.org/10.1007/978-3-031-60477-5_4

correctly. A substantial body of research also points to the absence of mental models as a critical factor contributing to reduced human performance in automated conditions. When confronted with unusual road conditions, drivers often lack an understanding of why an automated driving system behaves the way it does or anticipates certain actions. Therefore, possessing a correct mental model of AV systems is considered essential for drivers (Merriman, Plant et al. 2021a).

The widely accepted definition of a mental model is "the mechanism by which users describe the purpose and form of a system, interpret system functions and observed system states, and predict future system states" (Boelhouwer, van den Beukel et al. 2019). Mental models can influence drivers' system cognition (Merriman, Revell et al. 2023), trust and reliance (Boelhouwer, van den Beukel et al. 2019, Kraus, Scholz et al. 2020, Merriman, Revell et al. 2023), as well as attention and risk perception (Barg-Walkow and Rogers 2016, Merriman, Revell et al. 2023), thereby impacting drivers' performance in automated driving. The ability to have a clear mental model that accurately describes the AV system is considered essential for drivers (Merriman, Plant et al. 2021b), and how to establish such mental models is currently a primary focus of research.

Mental models are formed through training and experience (Merriman, Plant et al. 2021a). Training can effectively explain the fundamental logic of AV system to drivers and assist them in constructing appropriate mental models (Forster, Hergeth et al. 2019, Ebnali, Kian et al. 2020). As the most frequently encountered training material in real world, the mental model formed by reading the user manual can develop into a correct mental model which would fade over time if practical driving experience was lacking (Beggiato, Pereira et al. 2015). Solely relying on user manuals was insufficient as an effective teaching method for future automated driving systems (Boelhouwer, van den Beukel et al. 2019). Therefore, researchers are exploring other more effective training methods. Sportillo, Paljic et al. (2018) initially studied the driving training effectiveness of user manuals, fixed-screen driving simulators, and virtual-reality (VR) driving simulators. The results indicated that both fixed-screen driving simulators and VR driving simulators significantly improved drivers' takeover response speed, with drivers reporting better results for VR training. Furthermore, Sportillo, Paljic et al. (2019)also compared different training methods, such as video, augmented reality (AR), and VR, demonstrating that interactive training (AR and VR) could better enhance drivers' takeover performance. Results from Ebnali, Lamb et al. (2021) also indicated that VR, especially high-fidelity VR, positively influenced drivers' takeover performance and trust.

Research related to experience can be categorized into two main types. One focuses on the impact of automated driving experience on the formation of mental models, achieved through the design of training programs with different content. Simple trial-and-error exercises are not sufficient for drivers to form effective mental models (Beggiato and Krems 2013). Experiencing the limitations of AV systems can help drivers better understand the system's constraints (Teoh 2020). Merriman also points out that drivers need to understand what they should see from the environment and anticipate potential dangers (Merriman, Plant et al. 2021a). The other type of researches emphasizes the drivers' overall experience types before training, known as prior experiences. Prior experiences shape the driver's initial mental model, which would further update through practical experiences. If the mental model aligns with experience, the mental model will steadily grow (Beggiato and Krems 2013). The impact of prior experience with the use of AV systems on training can be both positive (Zeeb, Buchner et al. 2016) and negative (Lubkowski, Lewis et al. 2021a). Performance differences exist between users with experience and novices after training (Krampell, Solis-Marcos et al. 2020). However, there is currently limited research that combines prior experience and training for investigation.

This article aims to assess the mental models of drivers through various physiological and driving features. It analyzes and compares the impact of training methods on the mental models of drivers with different experience types. The goal is to determine the differences and identify the optimal training method for future automated driving.

2 Method

2.1 Participants

This experiment recruited 72 participants with three different experience types: infrequent drivers, frequent drivers with no exposure to AV systems, and frequent drivers with exposure to AV systems, with 24 participants in each group. These participants were assigned to one of four different knowledge transfer methods, ensuring an equal distribution of participants with different experience types across each knowledge transfer methods. All participants possessed a valid driver's license, with those having a total driving mileage of less than 1200 km classified as infrequent drivers and those with a total driving mileage exceeding 1200 km classified as frequent drivers.

2.2 Scenarios

Two groups of identical scenarios, labelled as A and B, were developed for training and testing. Each group consists of three caution scenarios and three takeover scenarios. Caution scenarios were utilized to analyze participants' eye-tracking behavior, using non-driving related tasks (NDRT) to assess drivers' attention and mental models. Meanwhile, takeover scenarios were used to analyze takeover actions, including takeover reactions, driving safety and stability. In both types of scenarios, the system issued a warning 7 s in advance, requiring the driver to decide whether to initiate a takeover. After each driving scenario, the system prompted the driver to switch to autonomous driving mode. Caution scenarios were extracted from the Tesla owner manual's sections related to limitations while takeover scenarios were extracted from warnings. Some limitations that did not explicitly discourage the use of AV systems were modified to serve as caution scenarios by introducing conditions favoring the use of AV systems or reducing the impact of environmental.

Safe Scenarios. The formal experiment driving covers a total distance of 25 km, with segments outside of caution and takeover scenarios being considered as safe scenarios.

Caution Scenarios. In caution scenarios, drivers were required to maintain heightened attention to ensure the proper functioning of AV system and prevent accidents. Takeover was not necessary unless there were abnormal conditions such as veering off the road.

1. Abnormal Road Surface with Lane Lines: a 500m stretch of dirt (Group A) or sandy (Group B) road surface was encountered, with no other vehicles on the road.

2. Small Vehicle Ahead: regular motorcycle (Group A) or racing motorcycle (Group B) appeared in front of the road. While driving at a speed of 80 km/h, if no takeover was initiated, AV system would maintain a 100m following distance within the same lane. After 500 m, the motorcycle would change lanes and decelerate.
3. Weather: foggy weather (Group A) or rainy weather (Group B) was encountered while driving, reaching its maximum severity after 100m of driving. The severity of weather factors was lower than the "Not Suitable for Use" restrictions mentioned in the Tesla owner manual.

Takeover Scenarios. In takeover scenarios, drivers were required to immediately take control of the vehicle. Failure to do so would result in a collision during the training. Although no collision occurred during the formal experiment for data consistency, participants were informed of the consequences of not taking over.

1. Unrecognizable Traffic Signal: obstructed traffic signal (Group A) or blinking traffic signal (Group B) appeared at a crossroads ahead while the vehicle was traveling at a speed of 80 km/h.
2. Curve: sharp curve (Group A) or multiple curves (Group B) appeared ahead while the vehicle entered the curve while maintaining a speed of 80 km/h.
3. Sudden Obstacle: the vehicle ahead deviated from its lane, and a red roadside marker appeared in the same lane (Group A) or a stationary large vehicle appeared on the side, with a pedestrian walking out from the front of the vehicle (Group B). If participants used group A scenarios for training, they would encounter group B scenarios during the formal experiment, and vice versa.

2.3 Training

Non-training Group. The non-training group received a simple introduction to the AV system, consisting of two parts: sensors' function and position, and the system's functions and principles. This section did not include any content related to scenarios, limitations of AV systems or driver operating requirements. It only provided basic knowledge about hardware and functionality, serving as a blank control group. The other three experimental groups were also required to read the same system introduction.
Owner Manual Training Group. Owner manual training was developed by PowerPoint software and includes information on the reliability and limitations of AV systems in various typical scenarios (Fig. 1). For each different scenario, participants were presented with illustrative images of the respective scenario, accompanied by textual descriptions of real road conditions, system limitations, recommended actions, and potential consequences.
Video Training Group. The content in the video training group was the same as that in owner manual training group. Participants first watched a video of an AV driving in typical scenarios (A or B) and were asked to make a takeover choice (Fig. 1). Depending on the choice made by participants, they were shown a corresponding video of the consequences of their choice, along with textual descriptions of system limitations, recommended actions, and potential consequences. If the participant's choice differed from the scenario type, they are required to re-watch the video for the same scenario again. If the choice aligns with the scenario type, they can proceed to the next scenario.

Simulator Training Group. The simulator training was developed by UC-Win/Road and included the same content as the owner manual training (Fig. 1). Participants in this group would drive a car equipped with the AV system on the driving simulator and experience typical scenarios (A or B). After each scenario occurred, the system prompts participants to make a takeover choice. If the participant made an incorrect choice or accidents occurred (e.g., collision), the system would provide an explanation of the error and the correct choice, and the participant must complete the scenario again after reading the explanation. If the choice is correct, the participant would receive a textual explanation with the same content, and they could choose to proceed to the next scenario after reading it.

Fig. 1. Example of training methods: Owner manual training (L), video training (M), simulator training (R).

2.4 Apparatus

The experiment utilized UC-Win/road for driving scenario construction, presented on three 24-in. screens. The simulator provided drivers with a 100-degree field of view and could collect driving data. Control of the simulated vehicle was achieved through a Logitech steering wheel and pedal controller.

2.5 Questionnaires

Throughout the experiment, participants will be required to complete 4 questionnaires, with 4 of them being reused.

Questionnaire 1: Basic Information. This questionnaire collects basic information about drivers, including age, gender, years of driving experience, total driving mileage, and experience with autonomous driving assistance systems.

Questionnaire 2: Mental Model. Developed based on Beggiato, Pereira et al. (2015) mental model questionnaire, this questionnaire consists of 14 scenarios. Participants rated their level of tension for each of the 14 statements on a scale from 1 "safe" to 4 "alert" and up to 7 "dangerous".

Questionnaire 3: Trust in Autonomous Driving Assistance Systems. This questionnaire is a single-dimensional trust questionnaire with 12 items (Beggiato and Krems 2013). Participants are required to rate their agreement with 12 descriptions related to AV systems based on their own experiences and understanding, ranging from 1 "completely disagree" to 7 "completely agree".

Questionnaire 4: Acceptance Five-Point Likert Scale. This questionnaire consists of 9 items (Beggiato and Krems 2013) and employs a five-point Likert scale to assess acceptance in two dimensions, usefulness and satisfaction.

2.6 Experimental Procedure

Before experiment, participants were required to submit information about their driving mileage and experience with AV systems online. If qualified, participants were randomly assigned to one of the four training groups, and the order of training and experimental scenarios was also randomly but equally determined. Upon arrival at the laboratory, participants sequentially completed Questionnaires 1 to 4. Participants were then instructed to practice basic operations on the driving simulator, including fundamental driving maneuvers, NDRT operations, and the activation and deactivation of the AV system. During NDRT operation practice, participants in all groups read a brief introduction to the AV system. After completing the basic operation practice, the non-training group proceeds directly to complete Questionnaire 2 again and then begins the formal experiment. The other three groups engaged in training before completing Questionnaire 2 and the formal experiment. The average learning time for the video training group and simulator training group was 10 min. The owner manual training group was allotted the same amount of time to read the materials, with most participants completing their reading within 3–5 min. During the formal experiment driving, participants were instructed to perform NDRT at any time while ensuring driving safety. NDRT involved watching a video segment about giant pandas. Questionnaire 2–4 were taken after the formal experiment driving.

2.7 Dependent Variables

Questionnaire Data. Mental Model Deviation Score: Calculated as the absolute difference between the score obtained for each scenario and the baseline score (1 for safe scenario, 4 for alert scenario, and 7 for takeover scenario). The mean score represents the quality of the mental model. A smaller mean score indicates a more accurate mental model, while a larger mean score indicates a less accurate mental model.

Trust: The mean score derived from 12 items, ranging from 1 to 7, with higher scores indicating greater trust.

Acceptance: The mean score calculated from 9 items, ranging from 1 to 5, with higher scores indicating greater acceptance.

Driving Data. Takeover Type: For the three takeover scenarios, calculate the frequency of each subject's takeovers and non-takeovers before the alert is issued.

Takeover Time: The time interval from the issuance of the prompt to when the driver initiates a response (e.g., steering wheel deviation of 10° or braking threshold exceeding 5%).

Road Fixation Time: The time interval from the issuance of the prompt to the first fixation of the driver's gaze on the road.

Driving Safety and Stability Metrics: During the manual driving period following takeover, metrics such as average speed, speed standard deviation, maximum longitudinal deceleration, maximum lateral acceleration, mean road deviation, and road deviation standard deviation.

3 Results

First, we checked whether all data met the assumptions of normality and homoscedasticity for the dependent variables. For data that met the assumptions, we conducted ANOVA to analyze the significance of the independent variables and their interaction effects. For data that did not meet the assumptions, we used the Kruskal-Wallis test. Bonferroni and Mann-Whitney tests were employed for post hoc analysis to determine the effects of the teaching. The significance level was set at 0.05. Additionally, Poisson regression was used only for the takeover type data for significance analysis.

3.1 Questionnaire Data Analysis

Mental Model Questionnaire. As results shown in Table 1. Before training, there were no significant differences in mental models among the groups. After training, the training methods showed significant main effects for the total scores of mental model ($F(3, 71) = 4.43$, $p = 0.007$) and driving experience also showed significant main effects ($F(2, 71) = 3.84$, $p = 0.003$). After driving experiment, only training methods exhibited significant main effects for the total scores of mental models ($F(3, 71) = 6.13$, $p = 0.001$). For different types of scenarios, training methods showed significant main effects for mental models in caution scenarios and takeover scenarios after training and driving (Fig. 2–3.). However, only driving experience showed significant main effects for mental models in caution scenarios after driving (Fig. 3).

Acceptance and Trust. Table 2 shows that there was a significant interaction between training method and driving experience for trust among the groups only after driving ($F(6,65) = 3.13$, $p = 0.01$). Post hoc tests indicated that, for the group with no experience, trust in the simulator training group was significantly higher than the video training group ($p = 0.04$). For frequent drivers with exposure to AVs, the differences between the owner manual training and the simulator training compared to the non-training group just missed the predefined level of statistical significance ($p = 0.08$).

Table 1. Summary of the significant and interaction of training methods and driving experience types on drivers' mental models

Time	Scenarios	Training methods	Driving experience	Interaction
Pre-transfer	Safe Scenarios	H(3,68) = 0.67, p = 0.88	H(2,69) = 1.08, p = 0.58	H(6,66) = 11.68, p = 0.07
	Caution Scenarios	H(3,68) = 7.29, p = 0.06	H(2,69) = 0.36, p = 0.83	H(6,66) = 2.55, p = 0.86
	Takeover Scenarios	F(3,68) = 0.41, p = 0.74	F(2,69) = 0.13, p = 0.88	F(6,66) = 0.56, p = 0.76
	Total	F(3,68) = 0.76, p = 0.52	F(2,69) = 0.40, p = 0.68	F(6,66) = 0.66, p = 0.68
Post-transfer	Safe Scenarios	H(3,68) = 3.15, p = 0.37	H(2,69) = 0.24, p = 0.89	H(6,66) = 6.34, p = 0.39
	Caution Scenarios	**F(3,68) = 3.07, p = 0.03**	F(2,69) = 1.84, p = 0.17	F(6,66) = 1.45, p = 0.21
	Takeover Scenarios	**F(3,68) = 6.78, p < .001**	**F(2,69) = 3.10, p = 0.052**	F(6,66) = 0.47, p = 0.83
	Total	**F(3,68) = 4.43, p = .007**	**F(2,69) = 3.84, p = .003**	F(6,66) = 0.80, p = 0.57
Post-driving	Safe Scenarios	H(3,68) = 4.10, p = 0.25	H(2,69) = 1.45, p = 0.48	H(6,66) = 5.30, p = 0.51
	Caution Scenarios	**F(3,68) = 3.66, p = 0.02**	**F(2,69) = 4.56, p = 0.01**	F(6,66) = 1.52, p = 0.19
	Takeover Scenarios	**F(3,68) = 7.09, p < .001**	F(2,69) = 0.23, p = 0.80	F(6,66) = 0.27, p = 0.95
	Total	**F(3,68) = 6.13, p = .001**	F(2,69) = 1.27, p = 0.29	F(6,66) = 0.52, p = 0.79

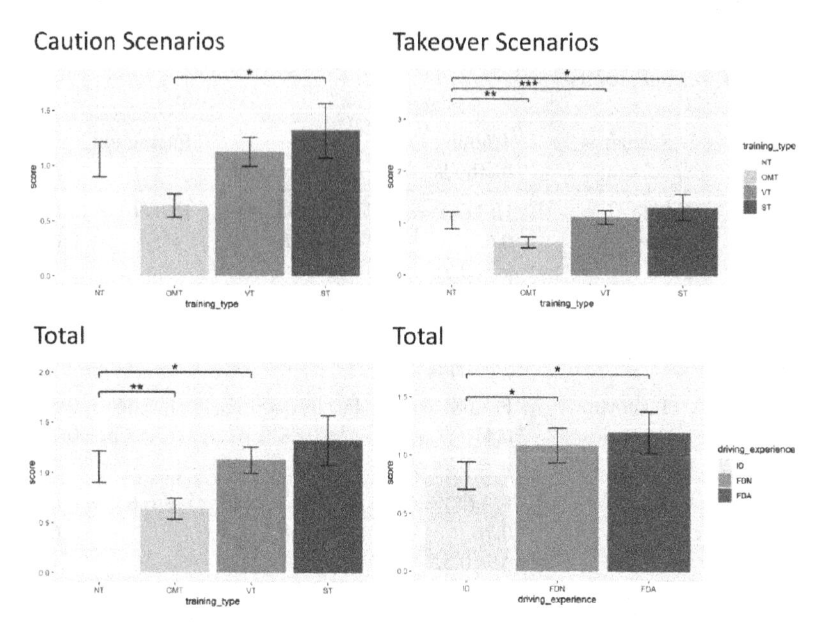

Fig. 2. Post-hoc contrasts of post-transfer mental model between groups.

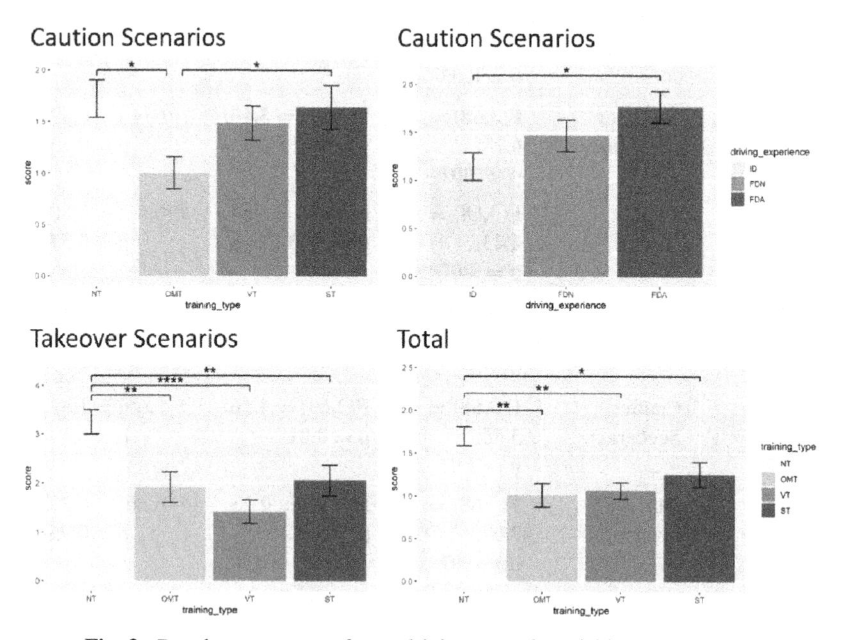

Fig. 3. Post-hoc contrasts of post-driving mental model between groups.

Table 2. Summary of the significant and interaction of training methods and driving experience types on drivers' acceptance and trust

Variable	Time	Training methods	Driving experience	Interaction
Acceptance	Pre-training	F(3,68) = 0.69, p = 0.56	F(2,69) = 0.47, p = 0.63	F(6,65) = 0.70, p = 0.65
	Post-driving	F(3,68) = 0.14, p = 0.93	F(2,69) = 1.32, p = 0.28	F(6,65) = 0.44, p = 0.85
Trust	Pre-training	F(3,68) = 0.12, p = 0.74	F(2,69) = 0.22, p = 0.80	F(6,65) = 1.62, p = 0.16
	Post-driving	F(3,68) = 0.41, p = 0.74	F(2,69) = 0.37, p = 0.69	**F(6,65) = 3.13, p = 0.01**

3.2 Driving Data Analysis

In the three takeover scenarios, takeover actions were categorized into three groups: takeover before the warning, takeover after the warning, and non-takeover. All metrics, except for road fixation time were analyzed only for participants who performed takeovers after the warning.

As results shown in Table 3. Among all the driving data, only training method showed a significant difference in road fixation time ($H(3,68) = 7.975$, $p = 0.047$).

Table 3. Summary of the significant and interaction of training methods and driving experience types on driving data

Variable	Training methods	Driving experience	Interaction
Takeover Time	F(3,63) = 0.999, p = 0.401	F(2,64) = 0.591, p = 0.557	F(6,60) = 1.108, p = 0.370
Road Fixation Time	**H(3,68) = 7.975, p = 0.047**	H(2,69) = 2.731, p = 0.255	H(6,65) = 3.093, p = 0.797
Longitudinal Speed (Average)	H(3,57) = 1.661, p = 0.646	H(2,58) = 2.762, p = 0.251	H(6,54) = 4.263, p = 0.641
Longitudinal Speed (SD)	H(3,57) = 1.356, p = 0.716	H(2,58) = 5.041, p = 0.080	H(6,54) = 4.033, p = 0.672
Max Longitudinal Acceleration	F(3,57) = 1.251, p = 0.302	F(2,58) = 2.121, p = 0.131	F(6,54) = 1.424, p = 0.226
Max Latitudinal Acceleration	H(3,57) = 1.038, p = 0.792	H(2,58) = 5.311, p = 0.070	H(6,54) = 5.252, p = 0.512
Offset From Lane Center (Average)	H(3,57) = 4.313, p = 0.230	H(2,58) = 1.673, p = 0.433	H(6,54) = 5.074, p = 0.534
Offset From Lane Center (SD)	H(3,57) = 1.055, p = 0.788	H(2,58) = 0.283, p = 0.868	H(6,54) = 4.223, p = 0.647

4 Discussion

In this study, we investigated the impact of training methods on the mental models, takeover performance, and eye-tracking behaviors of drivers with different driving experience. We also explored the interactions between these factors and their influence on trust and acceptance.

4.1 Questionnaire Data

The results of the mental model questionnaire data showed that infrequent drivers' mental model became closer to the predefined correct mental model after training, but the simulated driving experience weakened the differences between different post-training driving experience. All three training methods significantly improved the correctness of mental model, with simulated driving experience further increasing the significance of these differences. Since there were no significant initial differences in mental models among the participants, it can be concluded that infrequent drivers' mental models are more susceptible to the influence of training, but this influence becomes more similar to frequent drivers as driving experience increases. This is consistent with previous research findings that mental model tended to converge to a similar level with increased driving experience after training (Beggiato and Krems 2013). On the other hand, post-training driving experience influences drivers' mental model and further amplifies the existing differences. For takeover scenarios, the effects on participants' mental model after training are more significant. However, for caution scenarios, owner manual training was significantly better than non-training and simulator training, indicating that owner manual is still necessary under certain conditions (Detjen, Degenhart et al. 2021).

4.2 Driving Data

Video training significantly increased the frequency of drivers' frequency of takeover before the warning, which may primarily reflect the improvement in road attentiveness due to video training. For participants who completed takeover after the warning, most of the takeover performance data showed no significant differences. This result is consistent with previous research, indicating that drivers' mental models have almost no impact on preparing for and executing obvious reflex actions (Pampel, Jamson et al. 2018). However, it is worth noting that although there were no significant differences in takeover time, simulator training still significantly extended the road fixation time. This suggests that simulator training may reduce the time it takes for participants to understand the scenario and make takeover decisions, but it also slows down the speed at which participants return their attention to the road (Korber, Baseler et al. 2018, Detjen, Degenhart et al. 2021, Merriman, Plant et al. 2021a).

4.3 Limitations

This study has several limitations that should be noted. First, all participants were recruited within Shenzhen University. While their driving experience types met the experimental requirements, the non-representativeness of the sample group could potentially affect the generalizability of the results. Second, due to time constraints, the study only explored the effects of single-session knowledge transfer on attributes such as the drivers' mental model, without further investigating the impact of time and post-training driving experiences. In some experimental findings and previous research, these factors have been identified as important influences. Future research will aim to address these limitations and optimize the study design accordingly.

5 Conclusion

In summary, we conducted a simulated driving experiment to explore the effects of training methods on mental models, takeover performance, and eye-tracking behaviors of drivers with different levels of experience. The results indicate that both training methods and driving experience have a significant impact on the formation of drivers' mental models, and the types of scenarios and post-training driving experience further magnify the influence of training methods while mitigating the impact of driving experience. Video training is more effective than other training methods in improving drivers' road attentiveness, thereby enhancing driving safety. Finally, for drivers who have no exposure to AV systems, having more extensive driving experience may reduce their road attentiveness in hazardous road sections. These results highlight the differences in the effectiveness of AV system training for different types of drivers and emphasize the need to develop training programs tailored to driver types to improve their understanding of the capabilities, limitations, and requirements of Level 2 AV systems and enhance the safety of their use.

Acknowledgement. This work was supported by the Guangdong Basic and Applied Basic Research Foundation (grant number 2021A1515011610 and the Foundation of Shenzhen Science and Technology Innovation Committee (grant number JCYJ20210324100014040).

References

Barg-Walkow, L.H., Rogers, W.A.: The effect of incorrect reliability information on expectations, perceptions, and use of automation. Hum. Factors **58**(2), 242–260 (2016)

Beggiato, M., Krems, J.F.: The evolution of mental model, trust and acceptance of adaptive cruise control in relation to initial information. Transp. Res. F Traffic Psychol. Behav. **18**, 47–57 (2013)

Beggiato, M., Pereira, M., Petzoldt, T., Krems, J.: Learning and development of trust, acceptance and the mental model of ACC. A longitudinal on-road study. Transp. Res. F Traffic Psychol. Behav. **35**, 75–84 (2015)

Boelhouwer, A., van den Beukel, A.P., van der Voort, M.C., Martens, M.H.: Should I take over? Does system knowledge help drivers in making take-over decisions while driving a partially automated car? Transp. Res. F Traffic Psychol. Behav. **60**, 669–684 (2019)

Detjen, H., Degenhart, R.N., Schneegass, S., Geisler, S.: Supporting user onboarding in automated vehicles through multimodal augmented reality tutorials. Multimodal Technol. Interact. **5**(5) (2021)

Ebnali, M., Kian, C., Ebnali-Heidari, M., Mazloumi, A.: User experience in immersive VR-based serious game: an application in highly automated driving training. In: Stanton, N. (ed.) AHFE 2019. AISE, vol. 964, pp. 133–144 (2020). https://doi.org/10.1007/978-3-030-20503-4_12

Ebnali, M., Lamb, R., Fathi, R., Hulme, K.: Virtual reality tour for first-time users of highly automated cars: comparing the effects of virtual environments with different levels of interaction fidelity. Appl. Ergon. **90**, 103226 (2021)

Forster, Y., Hergeth, S., Naujoks, F., Krems, J., Keinath, A.: User education in automated driving: owner's manual and interactive tutorial support mental model formation and human-automation interaction. Information **10**(4) (2019)

Ganesh, M.I.: The ironies of autonomy. Humanit. Soc. Sci. Commun. **7**(1) (2020)

Korber, M., Baseler, E., Bengler, K.: Introduction matters: manipulating trust in automation and reliance in automated driving. Appl. Ergon. **66**, 18–31 (2018)

Krampell, M., Solis-Marcos, I., Hjalmdahl, M.: Driving automation state-of-mind: using training to instigate rapid mental model development. Appl. Ergon. **83**, 102986 (2020)

Kraus, J., Scholz, D., Stiegemeier, D., Baumann, M.: The more you know: trust dynamics and calibration in highly automated driving and the effects of take-overs, system malfunction, and system transparency. Hum. Factors **62**(5), 718–736 (2020)

Lubkowski, S.D., et al.: Driver trust in and training for advanced driver assistance systems in real-world driving. Transp. Res. F Traffic Psychol. Behav. **81**, 540–556 (2021)

Merriman, S.E., Plant, K.L., Revell, K.M.A., Stanton, N.A.: Challenges for automated vehicle driver training: a thematic analysis from manual and automated driving. Transp. Res. F Traffic Psychol. Behav. **76**, 238–268 (2021)

Merriman, S.E., Plant, K.L., Revell, K.M.A., Stanton, N.A.: What can we learn from automated vehicle collisions? A deductive thematic analysis of five automated vehicle collisions. Saf. Sci. **141** (2021)

Merriman, S.E., Revell, K.M.A., Plant, K.L.: Training for the safe activation of automated vehicles matters: revealing the benefits of online training to creating glaringly better mental models and behaviour. Appl. Ergon. **112**, 104057 (2023)

Pampel, S.M., Jamson, S.L., Hibberd, D.L., Barnard, Y.: Old habits die hard? The fragility of eco-driving mental models and why green driving behaviour is difficult to sustain. Transp. Res. F Traffic Psychol. Behav. **57**, 139–150 (2018)

Sportillo, D., Paljic, A., Ojeda, L.: Get ready for automated driving using virtual reality. Accid. Anal. Prev. **118**, 102–113 (2018)

Sportillo, D., Paljic, A., Ojeda, L.: On-road evaluation of autonomous driving training. In: 2019 14th ACM/IEEE International Conference on Human-Robot Interaction (HRI) (2019)

Teoh, E.R.: What's in a name? Drivers' perceptions of the use of five SAE Level 2 driving automation systems. J. Saf. Res. **72**, 145–151 (2020)

Zeeb, K., Buchner, A., Schrauf, M.: Is take-over time all that matters? The impact of visual-cognitive load on driver take-over quality after conditionally automated driving. Accid. Anal. Prev. **92**, 230–239 (2016)

Applying Theory of Planned Behavior to Explore the Safety Effects of Yellow Alert on Changeable Message Signs: Elicitation Interview Results

Pei Wang(✉) and Tingting Zhang

California PATH, University of California at Berkeley, Richmond, CA 94804, USA
peggywang@berkeley.edu

Abstract. The Yellow Alert program was proposed by the California Department of Transportation to establish a system designed to coordinate public alerts following a major injury or fatality producing hit-and-run collision. Sufficient vehicle information would be displayed on a changeable message sign (CMS) to either avert further harm or accelerate apprehension of the suspect. There are safety concerns with posting this "Alert", which may distract drivers if they look for the suspected vehicle, or if they change driving behaviors to examine surrounding vehicles. No prior studies have examined this topic. Therefore, the research team at California PATH program proposed a systematic human-factors study, aiming to evaluate the safety effects of Yellow Alert. First phase of this study, the elicitation interview, is reported here, which provides insights of drivers' potential responses to the Yellow Alert message and the underlying beliefs toward their responses framed with the model of Theory of Planned Behavior.

Keywords: Yellow Alert · Hit and Run · Changeable Message Sign · Elicitation Interview

1 Background

Hit-and-run crashes are a significant traffic safety concern in the US. Analysis found that both the rates of hit-and-run crashes and fatalities are increasing. Statistics from the Fatality Analysis Reporting System (FARS) reveal that: there were estimated 737,100 hit-and-run crashes in 2015, which resulted in 1,819 fatalities [1, 2]. At the state level, California has the highest number of hit-and-run fatalities compared with any other state in 2016. Three hundred and thirty-seven hit-and-run crashes resulted in at least one fatality in each crash in 2016 [3].

California Department of Transportation (Caltrans) proposed to conduct a two-year demonstration program to evaluate the effectiveness of displaying Yellow Alert messages on changeable message sign (CMS), with which the purpose was to establish a system designed to issue and coordinate public alerts following a major injury or fatality producing hit-and-run collision. With the proposed Yellow Alert CMS, there is sufficient vehicle information that could be displayed to either avert further harm or

H. Krömker (Ed.): HCII 2024, LNCS 14732, pp. 57–71, 2024.
https://doi.org/10.1007/978-3-031-60477-5_5

accelerate apprehension of the suspect. However, there has been no study that addresses the safety concerns with posting the Yellow Alert on CMS, which may distract drivers if they look for the subject vehicle, or if they change driving behaviors when examining and reporting a suspected vehicle, or even if they attempt to follow or stop the suspected vehicle.

2 Related Work and Research Questions

The proposed Yellow Alert program is like the well-established AMBERT Alert program, which is facilitated by a voluntary partnership between law-enforcement agencies and broadcasters [4]. The AMBER Alert program utilizes the Emergency Alert System and other means to notify the public about a missing child and the suspected abductor. However, very few studies have been conducted to evaluate the effectiveness of AMBER Alert on CMS. There was one driving simulation study conducted by Harder et al. [5], who explored the effectiveness of AMBER Alert. The tested message was "AMBER ALERT RED FORD TRUCK MN LIC# SLM 509". In the experiment, when participants passed the CMS displaying the AMBER Alert, the experimenter stopped the simulator and asked the participant "What was written on the Message Board over the road"? After the experiment, the 120 participants were assigned to one of four categories based on their Recall Scores: (1) Poor; (2) Fair; (3) Good; and (4) Excellent. Results showed that only 8.3% of the participants were included in the Excellent category, 51.7% were in the Good category, 31.7% were in the Fair category, and 8.3% were in the Poor category. Because of the low recall score for all participants, the authors suggested changing the content of the message.

Additionally, there have been several social science studies that investigated the direct benefits of AMBER Alert in recovering abducted children [6–8] However, impacts onto drivers' visual attention or driving behavior when reading or after reading the message was not evaluated in previous studies. In all, the effects of AMBER Alert on driving safety are still unclear. Therefore, further human-factors studies are needed to explore whether displaying the alert messages such as AMBER Alert or Yellow Alert on a CMS would impact on driving safety.

A systematic human-factors approach was planned to identify and evaluate the safety effects of Yellow Alert message on CMS, which includes an elicitation interview, an online survey, and a driving simulation experiment. We first conducted an elicitation interview aiming to explore drivers' possible behaviors of responding to Yellow Alert Message and to elicit salient behavioral, normative and control beliefs underlying the intention of their behaviors. More specifically, we wanted to answer the following questions through the elicitation interview:

1. What would be the potential responses when and after drivers see Yellow Alert message on a CMS?
2. What do drivers perceive to be advantages and disadvantages regarding responding to Yellow Alert?
3. Who would be the people or groups that the driver perceives to approve or disapprove of drivers' responding to Yellow Alert?
4. What factors facilitate or impede drivers' responding to Yellow Alert?

3 Methodology of the Elicitation Interview

3.1 Theory of Planned Behavior (TPB) Model

In this work, we adopt the Theory of Planned Behavior (TPB) model, depicted in Fig. 1, to develop the framework for the interview guide and relate the data gathered from the interview. We provide a brief overview of the TPB model here. The TPB model describes an individual's intention to carry out a behavior, which is based on a relationship between that person's attitudes, subjective norm, and perceived behavioral control towards that behavior. Attitude is beliefs about the likely consequences or other attributes of the behavior. Subjective norm is beliefs about the normative expectations of other people. Perceived behavior control is beliefs about the presence of factors that may further or hinder performance of the behavior [9].

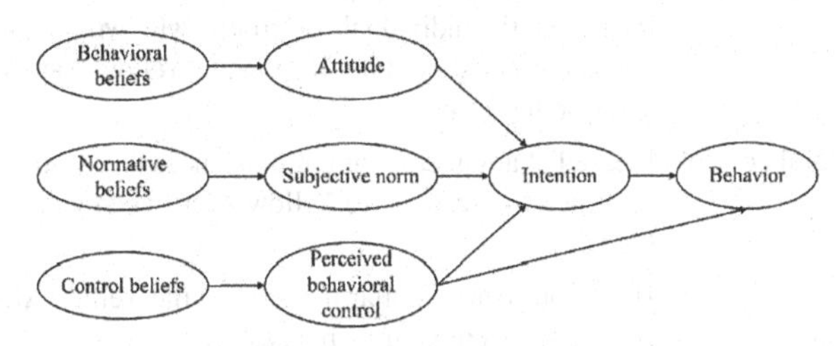

Fig. 1. Theory of Planned Behavior Model

3.2 Yellow Alert on CMS

The tested Yellow Alert message was "HIT AND RUN SUSP BLUE HONDA CIVIC CA PLATE 78LU469" (as shown in Fig. 2). The parameters were set based on the guideline for Model 500 CMS on CAMUTCD [3]. The font was Texas LED. The font size was 18 inches. The spacings were 25% character-spacing, 75% word-spacing, and 50% line-spacing.

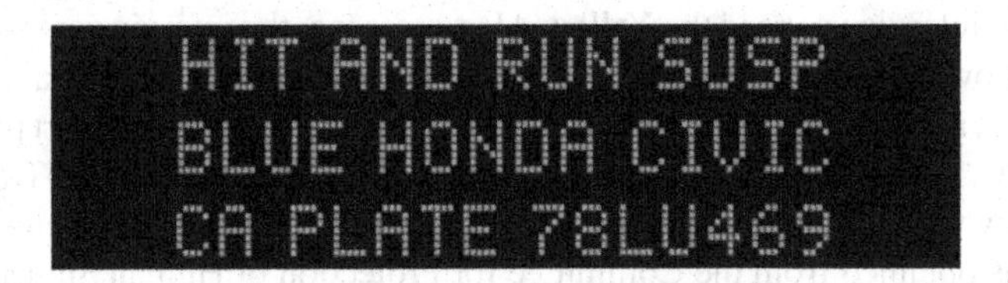

Fig. 2. Tested Yellow Alert Message

3.3 Interview Guide

An interview guide was developed following the guidelines of constructing a TPB survey [10]. Besides the main constructs of the TPB model, two more questions were added.

One question was about drivers' understanding of the Yellow Alert message. The other question was about drivers' experience of seeing and/or responding to a similar CMS message, which was AMBER Alert. The finalized interview guide is shown in Table 1.

Table 1. Elicitation interview guide

Constructs	Questions
Understanding	What does this sign mean to you?
Potential responses	What would be your responses if you see the Yellow Alert while driving on highway?
Attitudes	What do you see as the advantages / disadvantages of responding to the Yellow Alert while driving on highway?
Subjective norms	Please list the individuals or groups who would approve or think you should or should not respond to Yellow Alert when you drive on highway
Perceived behavioral control	Please list any factors or circumstances that would make it easy or difficult to respond to Yellow Alert when you drive on highway
Past behavior	Have you seen a similar message to the Yellow Alert message? If so, what were your responses?

3.4 Procedure

A semi-structured approach was used for the elicitation interview. All interviews were conducted individually, either in a conference room in Richmond Field Station of UC Berkeley or a quiet room (e.g., study room in a public library) according to interviewee's convenience. Each interviewee was firstly asked to fill out a demographic questionnaire. Secondly, they were introduced to the background and purpose of displaying Yellow Alert message on CMS and shown on a desktop monitor how the message would look like. After the interviewee read the Yellow Alert message, the interviewer started asking questions following the interview guide. On average, each interview took about 20 to 40 min. The interview process was recorded audially by using an app on a smart phone. After each interview, the audio file was uploaded onto YouTube in private mode. Transcripts of each interview were generated through YouTube thereafter. Approval of conducting the interview was obtained from the Committee for Protection of Human Subjects (CPHS) of UC Berkeley.

3.5 Sampling

In total, 26 Californian drivers were recruited to participate in the interview. The recruiting was designed to cover groups that could offer a broad range of different perceptive. All interviewees were above 18 years old and held valid California driver's licenses.

The participants, based on their demographic attributes, were classified into six different groups, including young drivers (25 years old and below), old drivers (65 years old and above), transportation professionals, non-native-English speaking drivers, drivers from rural areas, and general drivers. Four to six interviewees were included in each group. The interviewees' demographic information is summarized in Table 2. The mean value of age was 44.96 years old (SD = 17.8). The mean value of driving experience was 27.58 years (SD = 18.2). All interviewees had a high school or higher education. Among them, 15% reported that they had accidents, and 27% reported that they received traffic citations in the past five years.

Table 2. Interviewees' demographic information

Variables	Percentage	Variables	Percentage
Gender		Region	
Female	10 (38.5%)	Urban	22 (84.6%)
Male	16 (61.5%)	Rural	4 (15.4%)
Age		Native language	
Young (18–24)	6 (23.1%)	English	19 (73.1%)
Middle (25–65)	16 (61.5%)	Non-English	6 (23.1%)
Elder (>65)	4 (15.4%)	Not answered	1 (3.8%)
Education		Driving experience	
Middle school or lower	0 (0%)	<10 years	6 (23.1%)
High school	6 (23.1%)	10–30 years	10 (38.5%)
College	8 (30.8%)	30–50 years	5 (19.2%)
Postgraduate	12 (46.2%)	>50 years	5 (19.2%)
Accident reported in the past 5 years		Traffic ticket in the past 5 years	
Yes	4 (15.4%)	Yes	7 (26.9%)
No	22 (84.6%)	No	19 (73.1%)

4 Data Analysis

4.1 Indexing of the Transcripts

According to the approach of interview data analysis [11], the research team analyzed the data by taking the following steps. In the first step, researchers listened to the audio record once again, to get a sense of the interview, reviewed and revised the automatically generated transcript for each interview. In the second step, researchers divided all statements into different chunks and developed the content framework, following the questioning route. In the third step, researchers analyzed statements in each chunk, including highlighting and sorting quotes, making comparison between each case, and indexing. Different color codes were used for each index. In the fourth step, researchers extracted the quotes from each case and re-arranged them under the newly developed content framework. An example of the transcripts which were indexed as different attitude within the TPB framework is shown in Table 3.

Table 3. Example of indexing the transcripts

Line	Transcripts	Index
75	What do you see as disadvantages of responding to this?	
76	Distracted, you know it might distract drivers, you know they might be focused	3-13
77	on looking for those people. and even they find they might get a little too	
78	aggressive, you know either chasing or following this person.	3-14
79	Okay. distraction and aggressive behavior. anything else?	
80	If people are a little too aggressive and the person was a hit-and-run suspect	
81	who is a dangerous person, and they noticed these people are following, he/she	
82	could then hit and run the person following them or additional people.	3-10

5 Results

5.1 Understandability of the Yellow Alert Message

After reading the Yellow Alert message, interviewees were asked about its meaning. Results show that all 26 interviewees understood the message correctly. Specifically, they understood that a hit-and-run crash happened, with the color, make, model and license plate number of the suspect vehicle. They also understood that drivers were expected to report to the police if they saw the suspected vehicle. Interviewees were also asked to rate the easiness of understanding the message while driving on highway, on a scale of one to seven, by which one was extremely difficult and seven was extremely easy. The mean value of perceived easiness is 5.87 (STD = 1.41). Twenty-three interviewees rated it as 5 or higher. The other three interviewees rated it as 2, mainly because there were too many text characters and numbers to read while driving. One of the three interviewees explicitly commented that: "... especially the license plate number, it has seven digits on it". Besides too much information/texts/numbers, three other interviewees also mentioned that the abbreviation of "SUS" was not easy to interpret.

5.2 Potential Responses

The interviewees were asked "What would be your most possible responses when you see Yellow Alert CMS while driving on highway". Results are summarized in Table 4.

Table 4. Potential responses to Yellow Alert message

Potential response	Percentage	Exemplary verbatims
When I see the sign		
Read and try to remember the make, model, and a portion of the license plate number	12 (46.2%)	"...*Maybe I will remember only like two or three digits of the plate number.*"
Only read the color, make and model of the vehicle	7 (26.9%)	"*I would try to remember it because it's hit-and-run. I wouldn't pay any attention to the plate honestly because there's no way I'm going to remember that.*"
Read it and might try to write the license plate number down	4 (15.4%)	"*If the traffic was stopped you know it was very slow, then I might write it down or do a voice memo on my phone.*"
Just glance at it	3 (11.5%)	"*I will just read it a little bit. Then I understand it doesn't present to me and let it go. It doesn't involve me.*"
After passing the sign		
Keep an eye out for it while driving	16 (61.5%)	"*I would try to be aware to see if there is a blue Honda Civic in my field of vision.*"
Intentionally look for it	6 (23.1%)	"*I save that over in my head and then when I'm driving, I'll look around for a while for blue Honda Civic.*"
Do nothing if driving alone without other passengers	4 (15.4%)	"*I will just read it, that's pretty much of it.*"
If I see a vehicle which matches the make, model, and color (20 participants answered)		
Examine the plate number when it is safe	14 (70.0%)	"*If it's easy to change lane and I can easily follow that car to get a glance of the plate number. I'm not trying to just look into the plate by doing something dangerous.*"
Not move my car to examine	5 (25.0%)	"*No, no. It's too dangerous (to try to go behind the car).*"
Try everything to examine the plate number (may chase it if it speeds up)	1(5.0%)	"*Yeah, definitely. (If the car is driving very fast, I will try to chase it or follow it).*"
If I see the suspected vehicle		

(*continued*)

Table 4. (*continued*)

Potential response	Percentage	Exemplary verbatims
Use a hands-free system to call	8 (30.8%)	*"There's a microphone on the dashboard nearby. I just push the button."*
Pull over and call the police	6 (23.1%)	*"I would pull over and I would look up local police."*
Call the police when I'm not driving or let a passenger call	6 (23.1%)	*"I would probably never pull over. I think I will like even when I get home I try to make a phone call."*
Pick up my phone and call while driving	4 (15.4%)	*"To be honest with you I'd pick up my cell phone and use it even though probably it's not the best way to do it."*
Not call the police	2 (7.7%)	*"The police would catch up to this guy and arrest him. No, not a good thing to call the police."*

When I See the Sign. Among all interviewees, 23 (88.5%) of them said they would pay attention to the message if they saw the Yellow Alert CMS while driving on highway. The reason was that it's important to do so and hit-and-run is terrible. On the degree of reading detailed information on the sign, among the 23 who answered positively: Twelve interviewees (46.2%) stated that they would read the sign and try to remember a portion of the information including the first or last two or three digits of the license plate number. It would simply be too hard to remember them all. Another 7 (26.9%) interviewees said they would only read the make, model, and color of the suspect vehicle, without paying attention to the license plate number at all. Additionally, some of these seven interviewees mentioned that they would try to identify the suspect vehicle, which matches the description, and to get a hint of an accident on the vehicle, such as dent or smashed headlights. Besides reading and trying to remember, four interviewees (15.4%) reported that when the traffic was slow, they might try to retain the information on the sign, such as writing it down, making an audio memo or taking a picture of the sign. However, three interviewees (11.5%) out of 26 stated that the message was irrelevant to them.

After Passing the Sign. After passing the sign, 16 (61.5%) of interviewees said they would keep an eye out for the suspect vehicle. If a vehicle that matched the description came into their view, they would pay attention to it. Another six interviewees (23.1%) stated that they would look around and intentionally look for the suspected vehicle for a certain amount of time, which ranged from 30 s to 20 min. Four interviewees (15.4%) stated that they would not do anything if they were driving alone. But if there were passengers riding along, they might ask the passengers to look for the suspected vehicle.

If I See a Vehicle Matches the Description. Based on the answer to the previous questions, 22 interviewees would try to look for or pay attention to the suspected vehicle. These 22 interviewees were further asked whether they would try to examine the license plate number if a vehicle that matches the make, color and model of the suspected vehicle shows up within their vicinity. Twenty of them provided feedback. Among these 20 interviewees, (1) fourteen of them said they might try to get close to the vehicle in order to have a glance of the license plate number only if they felt it was safe to do so. To get close, they might change lanes or slow down their vehicles but not speed up. (2) Another 5 interviewees would not change the trajectory of their vehicles to check other vehicles' license plate number. (3) Only one of them said that he might chase the vehicle to check the plate number if the vehicle was speeding up.

If I Found the Suspected Vehicle. When seeing a vehicle which was likely to be the suspected vehicle, most of the interviewees would try to call the police, although in various ways. Out of the 26 interviewees: (1) eight interviewees (30.8%) would pull over and call the police. (2) Six interviewees (23.1%) said they would use their hands-free system, such as Bluetooth or CarPlay, to call the police while driving. (3) Another six interviewees (23.1%) would call the police when they are at home or let a passenger make the call. (4) Four interviewees (15.4%) said they would pick up the phone and call the police while driving. One of them explained that "I know I am not supposed to use the phone while driving. But I have to follow the vehicle while talking with the police. They will forgive me for this." (5) Two interviewees (7.7%) would not call the police.

5.3 Behavioral Beliefs

Behavioral beliefs regarding the perceived advantages and disadvantages of responding to Yellow Alert message are summarized in Table 5.

The most frequently identified advantage (n = 17, 65.4%) was "to help with catching the suspect of hit-and-run", which corresponds to the purpose of establishing the Yellow Alert program. Other advantages were also identified, such as helping the victim of hit-and-run (n = 9, 34.6%), preventing other drivers from committing hit-and-run (n = 8, 30.8%), and getting the dangerous driver (the suspect) off the road (n = 6, 23.1%). Seven interviewees (26.9%) said they felt satisfied because it was the right thing to do and it was their civil duty. Other advantages which were identified by less than 20% of the interviewees include: (1) it could remind other drivers of driving safely and (2) it was good karma, et al.

The disadvantage identified by most participants was driving distraction. Some interviewees were concerned that they might get themselves into an accident if they were to try to look for the suspected vehicle. Several interviewees were also concerned that some drivers might go too far and drive aggressively, like chasing the suspect vehicle or trying to stop it. These driving behaviors were identified as extremely dangerous to themselves and to other drivers on the road. Another disadvantage of responding to Yellow Alert message was the possibility of getting hurt or revenged by the suspect. This might happen at a point if the suspect was to stop other drivers from investigating their vehicle plate number or making the phone call. Some other disadvantages identified by less than 20% of the participants include: (1) causing traffic congestion if many drivers try to slow down and read the message; (2) getting themselves into trouble if the police call them as a witness; and (3) wasting their own time.

Table 5. Behavioral Beliefs of Responding to Yellow Alert

Beliefs	Percentage	Exemplary verbatims
Advantages (multiple inputs from one interviewee)		
Apprehend the suspect	17 (65.4%)	*"Maybe they would catch the suspect."*
Help the victim	9 (34.6%)	*"If somebody got hurt from the hit-and-run and that they didn't have medical insurance, (responding) would help the victim of the accident."*
Reduce hit-and-run	8 (30.8%)	*"Other people won't try to hit and run in another situation."*
Civil duty/do the right thing / be a good person	7 (26.9%)	*"(It is) my civil duty, just trying to do the right thing. Just satisfaction."*
Get the suspect off the road	6 (23.1%)	*"It will help in the long run to take these drivers off the road instead of being on the road again."*
Disadvantages (multiple inputs from one interviewee)		
Distract drivers	11 (42.3%)	*"If I'm trying to look for (it), I could see myself even potentially getting into a car accident."*
Waste the time of police if report falsely	8 (30.8%)	*"I could be wrong about the car that I see and then call the police. I could avert them to use their time productively."*
Get hurt or revenged by the suspect	5 (19.2%)	*"The person who committed the crime might find out who reported and come back to take revenge or..."*
Dangerous if somebody tries to pursue or stop the suspect	5 (19.2%)	*"I mean if somebody saw this car, he said oh I am going to follow them and I am going to catch them. They could be putting themselves in a dangerous situation."*

5.4 Normative Beliefs and Referents

The individuals or groups of people who would approve or disapprove responding to Yellow Alert message are summarized in Table 6.

Family (parents, wife/husband, kids) were perceived as people who would approve of responding to Yellow Alert message by eight interviewees (30.8%). They believed that their family would be proud of them if they responded to the Yellow Alert message. Friends or boyfriend/girlfriend were identified as people who would approve of responding to Yellow Alert message by five interviewees (19.2%). They would consider

Table 6. Normative referents of responding to Yellow Alert

Referents	Percentage	Exemplary verbatims
Family	9 (34.6%)	*"My family would be very proud of me (if I respond)."*
Nobody would disapprove	8 (30.8%)	*"Honestly, I don't think anybody would have an argument about that."*
Friends	5 (19.2%)	*"My girlfriend's opinion (matters). She will probably tell me to respond."* *"My friends too, would be proud of me."*
Police	5 (19.2%)	*"The majority of police officers would be very grateful."*

friends' opinion when deciding of responding or not responding. Another group of people who would approve of responding to Yellow Alert Message is the police who "put the message there", which was identified by five interviewees (19.2%).

Regarding the question of who would disapprove of responding to the Yellow Alert message, most interviewees thought responding to the message was the social norm and most people would approve it. Two interviewees reported that their family members would disapprove of it, especially the parents of young drivers. It is noted that all normative referents listed here were identified by less than 35% of the interviewees. One reason is that many interviewees thought they wouldn't be influenced by other people's opinion regarding responding to the Yellow Alert message. Nine interviewees (34.6%) explicitly expressed that they didn't care about other people's opinion in this case. One of them said: "I am strong-minded. Obviously, this is the right thing. I don't care what (other's people would say)."

5.5 Control Beliefs

Factors that would facilitate or prevent drivers from responding to Yellow Alert message are summarized in Table 7.

Bad weather conditions were the primary barrier for responding to Yellow Alert message. Eighteen interviewees (69.2%) reported that they would not pay much attention to the sign if it was raining, snowing or foggy. Further explanations include: (1) it would be hard to read the message and identify the vehicle because the visibility was poor, and (2) drivers should pay more attention to driving in bad weather conditions. Another factor identified by half of the interviewees (n = 13, 50.0%) was time urgency. They pointed out that it was much less likely for them to read or respond to Yellow Alert message if they were in a rush to somewhere or late for something. Drivers' mental status was another factor which might prevent them from responding to the Yellow Alert message. Ten interviewees (38.5%) commented that they would ignore the message if they had something else in their mind, worried about something or felt tired. Another barrier was the time of the day. Some interviewees stated that it would be hard for them to respond if the sun was hitting the sign in the afternoon, which would make it hard to read. Other interviewees mentioned that it was hard to read the message and see other

Table 7. Control beliefs of responding to Yellow Alert

Factors	Percentage	Exemplary verbatims
Bad weather conditions	18 (69.2%)	*"If it's raining, very much diminished the chance that I'm gonna pay much attention."*
In a rush	13 (50.0%)	*"Well if I were in a big hurry to get somewhere like I'm going to the airport to get a plane, (I might now pay attention)."*
Traffic*	12 (46.2%)	*"If there's a lot of traffic, it might be hard because if you might see them just like on the far other lane, it might be impossible to try."*
Mental status	10 (38.5%)	*"If I'm tired, already thinking about something else, (I won't pay attention)"*
Time of the day	7 (26.9%)	*"Well time of the day (matters), night time, it's harder to see the car. Early in the morning is another one (when it's harder to see)."*
Passenger*	7 (26.9%)	*"If somebody's in the car with me, and we're having a conversation, I'm less likely to pay attention to the sign."*
Cell phone status	6 (23.1%)	*"If there's no cell phone coverage or I forgot my phone, stuff like that, there's nothing anybody can do."*
Passenger*	7 (26.9%)	*"If I have someone else with me, that would be a safe situation (to have them help and respond)."*
Reporting app	7 (26.9%)	*"If I had my Bluetooth connected. (it would be easy to respond)."* *"If the reporting system is accurate (e.g. an app to report the car and send the location with a click), it is easy to access."*
Driving distance	6 (23.1%)	*"You weren't gonna be getting off the highway anytime soon, so you had more than enough time driving straight down."*
Traffic*	5 (19.2%)	*"If there was traffic and so I was kind of at a standstill, then (I have time to read and respond)."*

vehicles at night. Six interviewees (23.1%) mentioned that the status of the cell phone could be a barrier for them to respond to Yellow Alert message. They would not be able to respond if they didn't have a cell phone or the cell phone had a low level of battery power or didn't have signal.

The reporting system, such as a smart phone app, was a facilitator identified by seven interviewees (26.9%). Two of them said it would be easier if they had an app on the phone to report the suspected vehicle and their location with just one click. Another two interviewees mentioned that it would be helpful if they had their Bluetooth connected. Another one interviewee mentioned it would be easy if they had a camera to record the traffic including the suspected vehicle and the location. Longer driving distance would be another facilitator, as drivers were more likely to respond if they were on a long journey.

Two factors were identified with opposite effects of being a facilitator and being a barrier. One factor is the traffic. Twelve interviewees (46.2%) perceived heavy traffic as a barrier of responding to Yellow Alert message, since the driving itself would require more effort. Oppositely, another five interviewees (19.2%) perceived traffic as a facilitator for responding to Yellow Alert message. The reason was that in heavy traffic they would have more time to read and remember the message and investigate the vehicles in their vicinity. Two interviewees stated that if the traffic was slow enough, they would try to write the vehicle plate number down for later reference. Similar to traffic conditions, having a passenger was the other factor which was perceived as having opposite effects onto responding to Yellow Alert message. Seven interviewees (26.9%) commented that a passenger in the vehicle would be helpful as the passengers could read the message and look around. If the suspected vehicle was in the vicinity, the passenger could easily pick up the phone and call the police. But the same number of interviewees (n = 7, 26.9%) stated that they would be less likely to pay attention to the sign if there was a passenger in the car and they were having a conversation. Especially when they had kids in the car, the kids would a lot of their attention.

Some other facilitators identified by less than 20% of the interviewees include: (1) seeing repeated signs on the road; (2) indicating that the hit-and-run accident happened in a local area rather than a further-away area; and (3) familiarity with the road and the neighborhood. Some other barriers identified by less than 20% of the interviewees include: (1) bad vehicle condition; (2) the sign was blocked and not easy to read; and (3) knowing the hit-and-run incident happened not now but a while ago.

5.6 Past Behavior with AMBER Alert

At the end of the interview, interviewees were asked whether they had seen a similar sign – AMBER Alert message while driving on highway. If yes, what did they do when they saw it. Almost all interviewees had seen an AMBER Alert message on highway at least once. Nine interviewees (34.6%) said they didn't do anything when they saw the AMBER Alert message while driving on highway. One interviewee further explained that "I read it. But I never thought I could do anything about it." Seventeen interviewees (65.4%) reported they had tried to read and respond to the message. Among these seventeen interviewees, seven of them reported that after reading they looked around trying to find the suspect vehicle. Another seven of them reported that they kept driving to see if they happened to see the suspected vehicle. The results indicate that the percentage of interviewees who had responded to the AMBER Alert is relatively lower than the percentage of interviewees who would respond to the Yellow Alert.

6 Discussion and Conclusions

From the analysis of understandability, we can conclude that most drivers could easily understand the meaning of the Yellow Alert message, even though it was the first time they were introduced to this message during the interview. Similar to research findings about the AMBER Alert message [5], most drivers would have difficulty remembering all information displayed on the Yellow Alert CMS. The results from this interview phase of the study provide insights of drivers' potential responses to the Yellow Alert CMS when they drive on highway, which are summarized below. (1) Most of the interviewees (23 out of 26, 88.5%) would pay attention to the message and try to remember a portion of the license plate number. Only a few of them (4 out of 26, 15.4%) would try to retain the information for further reference by either writing it down or taking a picture of the sign. (2) After passing the sign, most of the interviewees (16 out of 26, 61.5%) would have an eye out for the suspected vehicle and at the same time keep driving. Several interviewees (6 out of 26, 23.1%) would intentionally look for the suspected vehicle for a certain amount of time. (3) If seeing a vehicle matches the description of the suspected vehicle, most interviewees (14 out of 20, 70.0%) would try to have a glance of the license plate but only when it is safe for them to do so. Only one interviewee (1 out of 20, 5%) would chase the vehicle to check the plate number if the suspected vehicle speeds up. (4) If the suspected vehicle was identified, a few of the interviewees (4 out of 26, 15.4%) would pick up the phone and call the police while driving. From the results, we could infer that displaying the Yellow Alert on a CMS would have impacts on most drivers' attention and driving behavior. The impact on a small portion of the drivers could be negative when they dramatically direct their attention to look for the suspected vehicle and alter their driving behavior to examine and report a suspected vehicle.

The results also provide insights into drivers' underlying beliefs towards potential responses. For the behavioral beliefs, most interviewees stated that the most-likely positive consequence of responding to the Yellow Alert message was to apprehend the suspect of hit-and-run, which is well aligned with the Caltrans' objective of establishing the Yellow Alert program. The most-likely negative consequence of responding to the message was driving distraction and potential involvement in an accident. For the normative beliefs, most interviewees thought responding to the message was the social norm and most of their family members or friends would approve of their responding. For the control beliefs, the three most frequently identified factors that would influence their responses include (1) adverse weather conditions; (2) time urgency; and (3) mental status. These beliefs further explained why drivers would or would not respond to the Yellow Alert message.

Due to the nature of the interview approach, limitations of this phase of the study are two-fold. Firstly, the number of participants in the elicitation interview is limited. The findings here provide insights of what are drivers' potential responses to the Yellow Alert message. But it could not be generalized to all driver population in California in terms of what percentage of the drivers are going to take the risky responses (e.g., change lane to exam the suspected vehicle, or chase the suspected vehicle if it speeds up, call and report to the police while driving). Therefore, based on findings from the elicitation interview, a survey questionnaire was developed to gather the data from a larger sample size who would better represent the driver population in California. Secondly, the potential responses revealed from the interview could be categorized as relatively safe or dangerous. There is no quantitative measurement for the impact of driving safety. To overcome this drawback, a driving simulator study was designed to measure drivers' visual attention and driving behavior change when the Yellow Alert CMS and the suspected vehicle are presented, in comparison with their visual attention and driver behavior without Yellow Alert CMS or suspected vehicle. The results from the second and third phases of the study are not included here, due to space limitation, and will be discussed and published separately.

Acknowledgments. The research team would like to thank Caltrans for funding this research project. Their support throughout this project is highly appreciated. The views and opinions expressed in this article are those of the authors and do not necessarily reflect the official policy or position of Caltrans and associated agencies.

Disclosure of Interests. The authors have no competing interests to declare that are relevant to the content of this article.

References

1. NHTSA: National Automotive Sampling System (NASS) General Estimates System (GES): Analytical User's Manual 1988–2015 (2016). https://crashstats.nhtsa.dot.gov/#/PublicationList/107
2. NHTSA: Fatal Accident Reporting System: Analytical User's Manual 1975–2017 (2018). https://crashstats.nhtsa.dot.gov/#/PublicationList/106
3. Benson, A., Arnold, L., Tefft, B., Horrey, W.J.: Hit-and-run crashes: prevalence, contributing factors and countermeasures. In: AAA Foundation for Traffic Safety (2018). https://aaafoundation.org/hit-and-run-crashes-prevalence-contributing-factors-and-countermeasures/
4. Caltrans: 2014 California Manual on Uniform Traffic Control Devices (CA MUTCD, Rev 4) (2014). https://dot.ca.gov/programs/traffic-operations/camutcd
5. Harder, K.A., Bloomfield, J., Chihak, B.J.: The effectiveness and safety of traffic and non-traffic related messages presented on changeable message signs (CMS). Minnesota Department of Transportation, Minnesota (2003). http://www.its.umn.edu/Publications/ResearchReports/reportdetail.html?id=703
6. Greer, J.D., Pan, P.-L., Flores, D., Collins, M.C.: Priming and source credibility effects on individual responses to amber and other mediated missing child alerts. Soc. Sci. J. **49**(3), 295–303 (2012)
7. Griffin, T., Miller, M.K., Hoppe, J., Rebideaux, A., Hammack, R.: A preliminary examination of AMBER Alert's effects. Crim. Justice Policy Rev. **18**(4), 378–394 (2007)
8. Griffin, T.: An empirical examination of AMBER Alert 'successes'. J. Crim. Justice **38**(5), 1053–1062 (2010)
9. Ajzen, I.: The theory of planned behaviour: reactions and reflections. Psychol. Health **26**(9), 1113–1127 (2011)
10. Ajzen, I.: Theory of planned behaviour questionnaire. Measurement Instrument Database for the Social Science (2013). http://www.midss.org/content/theory-planned-behaviour-questionnaire
11. Rabiee, F.: Focus-group interview and data analysis. Proc. Nutr. Soc. **63**(4), 655–660 (2004)

Comparative Study on the Effects of W-HUD and AR-HUD on Driver Behavior Based on Cognitive Load Theory

Huibo Xu, Xiangxuan Tang, Qilin Gu, and Jianrun Zhang(✉)

School of Mechanical Engineering, Southeast University, Nanjing 211189, China
zhangjr@seu.edu.cn

Abstract. Currently, W-HUD and AR-HUD are the two dominant HUDs. Although driving with HUD assistance improves safety and driver concentration, its effect on cognitive load is still unclear. This research aimed to methodically evaluate and scrutinize how W-HUD and AR-HUD influence the drivers' cognitive load and identify the most efficient HUD system for lessening the drivers' cognitive load. The experiment was designed to measure the subjects' behavioral performance on three different tasks: turning, altering speed, and identifying obstacles under the same road conditions. At the end of the experiment, the subjects filled in the NASA-TLX scale for each of the two HUDs for subjective evaluation. The results showed that drivers' cognitive load was lower when using the AR-HUD than the W-HUD and that drivers subjectively preferred to use the AR-HUD to assist driving. The results serve as a crucial guide for creating safer, more efficient in-vehicle information systems, aiding in wider application and upgrade iteration of HUD.

Keywords: W-HUD · AR-HUD · Cognitive Load

1 Introduction

With the escalation in the volume of data shown in a vehicle, the conventional center control screen falls short of presenting all driver-related information. Consequently, heads-up displays (HUDs) are extensively utilized in car cabin interiors to improve driving experiences. Head-up display (HUD) is an in-vehicle visual aid that displays driver assistance information such as current speed, direction navigation, system warnings, and the surrounding environment information in a reasonable superimposed manner in the field of view area in front of the driver [1]. Research by Jing et al. [2] endorses the effectiveness of HUD in diminishing visual distractions for drivers, enhancing the speed of gathering visual data, decreasing the frequency of eye movements from the road, and bolstering driving safety. With the progression of technology, the public's interest in driving has significantly increased. HUD has emerged as a crucial tool for optimizing human-machine interactions in future driving situations.

H. Krömker (Ed.): HCII 2024, LNCS 14732, pp. 72–81, 2024.
https://doi.org/10.1007/978-3-031-60477-5_6

Head-up displays (HUD) include combiner HUD (C-HUD), windshield HUD (W-HUD), and augmented reality HUD (AR-HUD), as shown in Fig. 1. C-HUD uses translucent resin imaging, while W-HUD and AR-HUD use optical technology to project information onto a vehicle's front windshield. AR-HUD enhances the integration of the projected information with the actual road conditions based on W-HUD. Medenica et al. [3] found that drivers using W-HUDs had longer periods of road concentration and experienced significantly less workload than those using C-HUDs, which implies that W-HUD provides better driving performance. Charissis et al. [4] found that vehicles using W-HUD had significantly fewer collisions than C-HUD in low visibility. Chauvin et al. [5] discovered that the implementation of an AR-HUD resulted in fewer mistakes and enhanced average driving speeds. Drivers subjectively perceived AR-HUD as more useful and easier to understand than C-HUD. To sum up, numerous research findings indicate that W-HUD and AR-HUD outperform C-HUD considerably. Furthermore, C-HUD has been gradually eliminated from the market due to its smaller projection capacity and decreased safety features. Consequently, this research focuses on W-HUD, presently a dominant player in the market, and AR-HUD, which is gradually gaining a larger market share, for a comparative analysis.

Fig. 1. C-HUD, W-HUD, and AR-HUD

Implementing HUD enhances both driving ease and safety. Nonetheless, given the clear intersection of HUD's display zone and the driver's visual range on the road, certain research worries that HUD might divert the driver's focus and heighten their mental effort [6]. Cognitive load characterizes the occupancy rate of brain resources under working conditions. Overburdened cognitive load may impair drivers' awareness of their surroundings, thereby impairing their capacity to identify and react to pertinent data in swiftly evolving road scenarios and leading to accidents caused by humans [7]. Most of the prior research has concentrated on the practicality of substituting the conventional C-HUD with W-HUD and AR-HUD, yet the impact of these technologies on the cognitive load of drivers remains insufficiently studied.

Numerous research efforts have been made to examine and assess the effectiveness of the developing AR-HUD in aiding driving. Research by Cheng et al. [8] gathered data on the eye movements of drivers and their response times to risks through five common risky driving situations day and night. Findings indicate that AR-HUD enhances drivers' awareness of hazardous areas and reduces their time to perceive and respond to perilous scenarios during nighttime driving. Research by Hwang et al. [9] investigated how drivers react and how their mental states alter in the face of unexpected road hurdles. Findings suggest that the AR-HUD system used in vehicles could alleviate stress or tension to some extent for drivers experiencing significant interpersonal anxiety. To summarize, most of the recent research has focused on the performance of W-HUD or AR-HUD in assisted driving, affirming their superiority over the traditional C-HUD. Nonetheless, there remains an absence of comparative research between W-HUD and AR-HUD. Therefore, it is meaningful to compare the effects of W-HUD and AR-HUD drivers' cognitive load.

The study aimed to determine which HUD is more effective in reducing driver cognitive load. On this basis, the advantages of HUD technology in improving driving safety and efficiency are further discussed. In this study, we evaluated the cognitive load of drivers using W-HUD and AR-HUD in combination with subjective drivers' operational performance evaluations. We investigated and compared the cognitive load of W-HUD and AR-HUD drivers performing different driving tasks. Driving tasks include three core display elements of the HUD display system and simulating actual road conditions that may occur in real-world driving situations, to evaluate the performance of HUD more broadly.

2 Methodology

2.1 Materials

In the experiments, there are three distinct HUD interface sets for various tasks, with each interface set comprising two sets, W-HUD and AR-HUD, and an empty array of interfaces lacking HUD as a prelude to the official experiment. Park and Im [10] found through extensive literature research that HUD's primary display elements include indicators of turn direction, current speed, system alerts, and road traffic data, among others. To minimize distractions from excessive information for drivers, this research established three separate experimental activities post-evaluation of the core components of the HUD display system: turning, altering speed, and identifying obstacles, and designed three groups of HUD interfaces accordingly. As to variable management, the experimental tasks took place under identical driving conditions, implying that each experimental material was crafted with a uniform backdrop image, differentiated solely in the HUD segment, and only the essential details required for the respective task were displayed. The interfaces of the W-HUD and AR-HUD experimental groups corresponding to the three experimental tasks are shown in Fig. 2, Fig. 3, Fig. 4. The experimental questionnaire used the NASA-TLX scale to subjectively evaluate the cognitive load of the subjects.

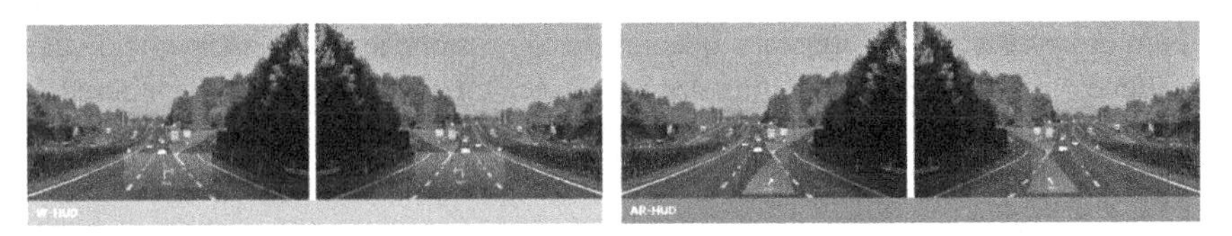

Fig. 2. Task 1-Turning

Fig. 3. Task 2-Altering Speed

Fig. 4. Task 3-Identifying Obstacles

2.2 Subjects

20 subjects were enrolled from school of mechanical engineering in Southeast University. Subjects included 10 males and 10 females between the ages of 22 and 31 years (M = 24.8), all of whom were right-handed and had normal vision or corrected vision. Each participant in the experiment possessed prior automotive driving experience but lacked HUD usage skills, and they were all briefed about the experiment's objectives before its commencement.

2.3 Experimental Equipment and Experimental Procedures

This research employed the W-HUD and AR-HUD interfaces as materials, assessing the cognitive load on participants using these HUDs via subjective judgment and behavioral analysis, followed by a comparative study. The essence of assessing cognitive load lies in the rational management and efficient use of cognitive resources, where too much cognitive load can result in reduced operational efficiency, which may lead to more serious human-caused accidents. "The primary methods for measuring cognitive load are subjective evaluation measures, behavioral performance measures, and physiological indicator measures [7]." Given that physiological indicators' measurement tools can disrupt driving patterns, this research did not employ physiological measurements to evaluate cognitive load. For the experiment's precision, the research utilized a mix of subjective evaluation and behavioral analysis to gauge the subjects' cognitive load. Evaluation of the behavioral measurement phase involved documenting the participants' reaction times and accuracy, while the subjective evaluation phase was determined by completing the NASA-TLX scale post-experiment. Figure 5 displays the research flow of the experimental research.

The experiments were conducted in a quiet environment with normal lighting conditions. The experimental equipment were laptop computers with screen size in the range of 13 to 17 inches. This experimental program is written by E-Prime software, which will record the time from the emergence of the experimental material to the participants' reaction and the accuracy rate of the reaction. Subjects are required to finish the three

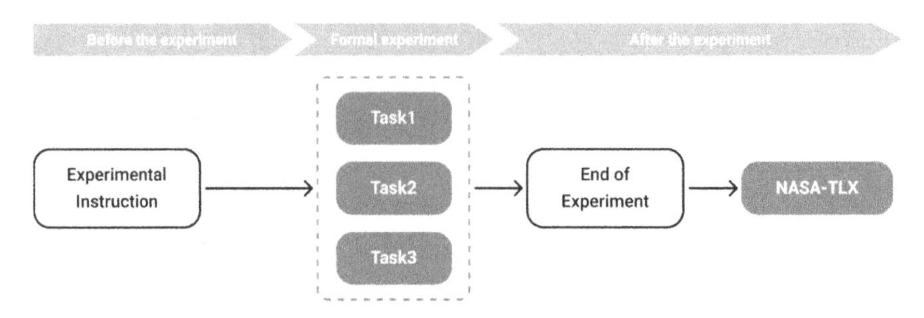

Fig. 5. Research Flow

experimental tasks: turning, altering speed, and identifying obstacles. Before each task, the experiment will be introduced by the corresponding blank group materials, and the subjects will know how to operate the experiment and press the space bar to enter the formal experiment. During the experiment, the W-HUD and AR-HUD interfaces will emerge randomly and cycle three times. Upon knowing the experimental content, participants are required to swiftly respond with the keyboard's " ← " and " → " arrows and the software will auto-play the subsequent image. After all experimental tasks were completed, subjects completed the NASA-TLX scale to evaluate W-HUD and AR-HUD, respectively. The specific experimental procedure was as follows.

Before the Experiment. Inform the subjects of the purpose, procedure, and operation of this experiment.

Formal Experiment. Subjects need to complete three independent experimental tasks.

Task 1-Turning. Subjects watch the image on the display and decide to turn left or right. Press "←" to turn left, and press "→" to turn right.

Task 2-Altering Speed. Subjects watch the image on the display and determine if the present vehicle needs to alter its speed by the speed limit indicators shown on the screen. Press "←" if it needs to change speed, and press "→" if it doesn't need to change speed.

Task 3-Identifying Obstacles. Subjects watch the image on the display and judge whether the current vehicle needs to avoid the obstacles. Press "←" if it needs to avoid obstacles, and press "→" if it doesn't need to.

After the Experiment. Subjects need to complete the NASA-TLX assessment scale for appraising W-HUD and AR-HUD, respectively.

3 Results

The correlation variables in the experiment were analyzed by using the statistical software SPSS. For reaction time, the normality of the data was tested by the Shapiro-Wilk test ($\alpha = 0.05$). After the test, it was found that the data of the three experimental tasks did not obey normal distribution ($P = 0.000, 0.024, 0.004, P < \alpha$), so the results were analyzed by using the independent samples Mann-Whitney test. For response accuracy, since it did not satisfy continuity, it was analyzed by comparing means and chi-square tests. For NASA-TLX scale scores, the Shapiro-Wilk test ($\alpha = 0.05$) revealed that the data did not obey normal distribution ($P = 0.030$, $P < \alpha$), therefore the results were analyzed using the independent samples Mann-Whitney test.

3.1 Reaction Time

Figure 6 shows the results of the reaction time in Task 1-Turning. The AR-HUD group's average reaction time ($M = 0.719$) is less than that of the W-HUD group ($M = 0.801$). According to the Mann-Whitney test, the median of AR-HUD and W-HUD in terms of reaction time in task 1 is 0.641/0.735. The P-value was 0.299 ($\alpha = 0.05$, $P > \alpha$), thus there was no significant difference between AR-HUD and W-HUD. The variance in Cohen's d-value stood at 0.562, indicating a moderate level of disparity (where 0.20, 0.50, and 0.80 represented small, medium, and large critical difference points.)

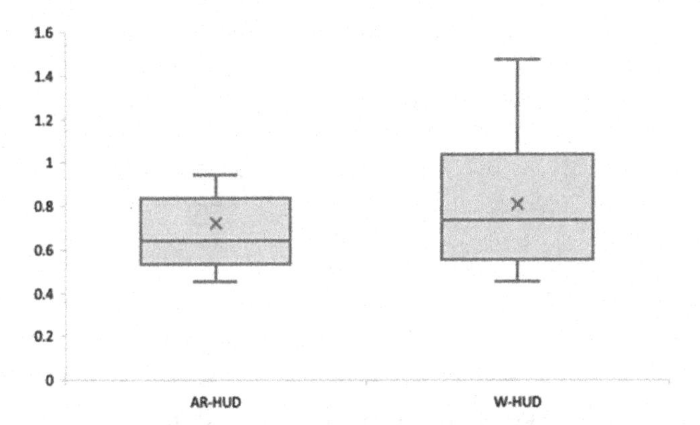

Fig. 6. The reaction time(s) of task 1

Figure 7 shows the results of the reaction time in Task 2-Altering speed. The AR-HUD group's average reaction time ($M = 1.216$) is less than that of the W-HUD group ($M = 1.491$). According to the Mann-Whitney test, the median of AR-HUD and W-HUD in terms of reaction time in task 2 is 1.094/1.337. The P-value was 0.042 ($\alpha = 0.05$, $P < \alpha$), thus there was significant difference between AR-HUD and W-HUD. The variance in Cohen's d-value stood at 0.803, indicating a significantly large disparity.

Figure 8 shows the results of the reaction time in Task 3-Identifying obstacles. The AR-HUD group's average reaction time ($M = 0.958$) is less than that of the W-HUD group ($M = 1.135$). According to Mann-Whitney test, the median of AR-HUD and W-HUD in terms of reaction time in task 3 is 0.939/1.115. The P-value of the test was 0.176 ($\alpha = 0.05$, $P > \alpha$), thus there was no significant difference between AR-HUD and W-HUD. The variance in Cohen's d-value stood at 0.563, indicating a moderate level of disparity.

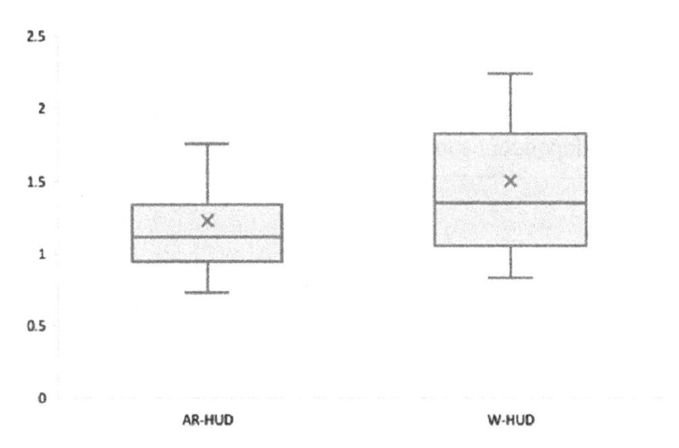

Fig. 7. The reaction time(s) of task 2

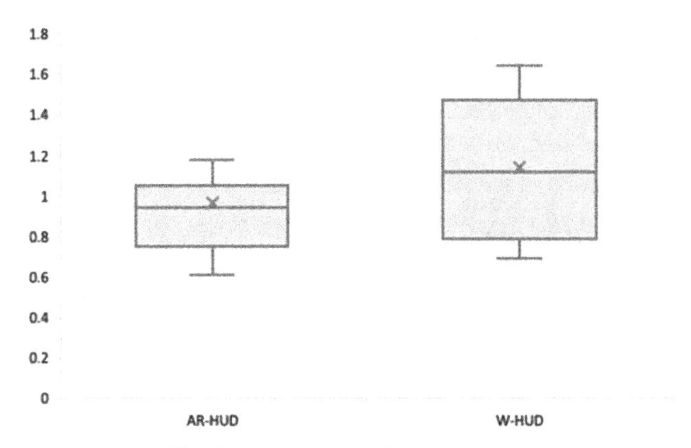

Fig. 8. The reaction time(s) of task 3

3.2 Reaction Accuracy

Table 1 displays the accuracy percentages of AR-HUD and W-HUD in experimental tasks. The chi-square test outcomes indicated no notable variance in accuracy rates between AR-HUD and W-HUD across the three tasks ($P = 0.311, 0.677, 0.292$, $\alpha = 0.05$, $P > \alpha$).

Table 1. Reaction accuracy percentages

HUD Group	Turning	Altering speed	Identifying obstacles
AR-HUD	100%	85%	85%
W-HUD	95%	80%	95%

3.3 NASA-TLX Scale

Figure 9 shows the results of NASA-TLX. According to the Mann-Whitney test, the median of AR-HUD and W-HUD on NASA-TLX scale scores were: 11.96/19.58. The P-value was 0.000 ($\alpha = 0.05$, $P < \alpha$), thus there was significant difference between AR-HUD and W-HUD on the cognitive load of subjects. The variance in Cohen's d-value stood at 2.135, indicating a significantly large disparity.

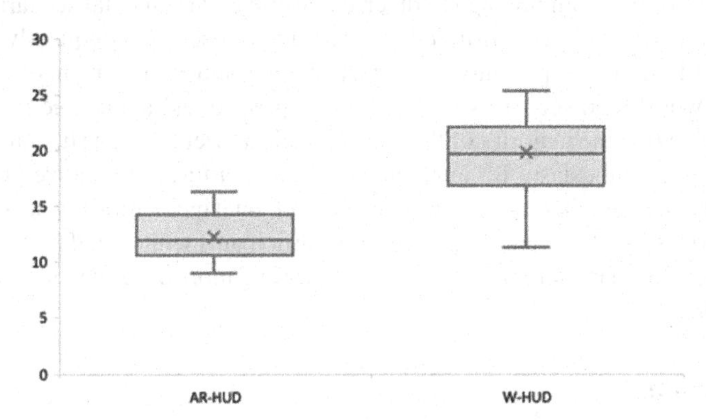

Fig. 9. NASA-TLX scale scores

4 Discussion

Overall, combining performance measures and subjective ratings, the experimental results show that drivers' cognitive load is lower when using an AR-HUD than when using a W-HUD. This implies that the AR-HUD provides better driving performance.

In terms of reaction times, the experimental results showed no significant difference in the reaction times between the two HUDs in tasks involving turning and identifying obstacles ($P = 0.299, 0.176$, $\alpha = 0.05$, $P > \alpha$). However, a significant difference was noted in the reaction times between the two HUDs in the altering speed task ($P = 0.042$, $\alpha = 0.05$, $P < \alpha$), likely because the AR-HUD incorporated navigation information into the real environment, enhancing the driver's comprehension of the actual environment.

Regarding the accurate response rate, the experimental findings indicate no significant difference in the correct response rates between AR-HUD and W-HUD across the three tasks ($P = 0.311, 0.677, 0.292$, $\alpha = 0.05$, $P > \alpha$). The high rate of correct responses suggests that both HUDs provide good and safe driver assistance. Notably, during tasks involving altering speed and dodging obstacles, the AR-HUD's accuracy rate was marginally less compared to the W-HUD. This may be because the visual center of the W-HUD is always in the driver's field of vision, but the visual center of the AR-HUD shifts to different objects as the task changes. The non-constancy of the visual center distracts the driver's attention and leads to misjudgment to some extent.

For subjective evaluation, the scores of the NASA-TLX scale showed that the cognitive load when using AR-HUD was significantly lower than that of W-HUD ($P =$

0.000, $\alpha = 0.05$, $P < \alpha$), suggesting that subjects subjectively preferred AR-HUD and believed that AR-HUD would make driving easier. This may be because AR-HUD is tightly integrated with road realities, providing a more intuitive navigation experience for drivers, and improving driving comfort.

There are limitations in the current study that could be addressed in future research. To circumvent the impact of varying road conditions on the outcomes of the experiments, the experimental tasks in this study were executed under identical road circumstances. Yet, it falls short of encompassing the diverse real-life environmental scenarios, including driving at night, the glare of sunlight, or climatic impacts. Consequently, upcoming studies ought to delve deeper into the impact of environmental elements by emulating diverse real-world road scenarios to derive more precise and applicable insights. Furthermore, given that the majority of the study's subjects were design students, they tend to exhibit greater enthusiasm for augmented reality as a form of creative presentation. Consequently, there might be a more pronounced personal inclination towards AR-HUD. Future research should contemplate increasing participants with varied professional histories and identity characteristics to offer a broader and more accurate view of assessing diverse HUDs.

5 Conclusions

The purpose of this study was to comprehensively assess the cognitive load of drivers using W-HUD and AR-HUD by examining driver performance during different driving tasks and combining this with the drivers' subjective ratings. This study aimed to determine which HUD systems are more effective at reducing cognitive load for drivers. Based on this criterion, we further explored which HUD technology performs more superiorly in improving driving safety and efficiency. Through the summary and selection of HUD core display elements, three experimental tasks were designed, namely turning, altering speed, and identifying obstacles under the same road conditions, and subjects were asked to respond accordingly. Response time and accuracy were recorded to assess performance, and subjects were asked to rate the two HUDs using the NASA-TLX scale.

As expected, the experimental results show that the cognitive load of drivers using AR-HUD is lower than that of those using W-HUD, and drivers are more likely to use AR-HUD-assisted driving subjectively. This paper's data examination and findings offer theoretical backing for HUD's evolution in future driving contexts and serve as a guide for the applications of W-HUD and AR-HUD in diverse situations. In future studies, diverse environmental conditions should be considered to simulate road conditions better suited to real-world driving scenarios for experimental use. Recruitment should include a wider range of participants with different occupational backgrounds and identity traits to ensure a comprehensive assessment of HUD.

References

1. Beck, D., Jung, J., Park, J., Park, W.: A study on user experience of automotive HUD systems: contexts of information use and user-perceived design improvement points. Int. J. Hum.-Comput. Interact. **35**(20), 1936–1946 (2019). https://doi.org/10.1080/10447318.2019.1587857

2. Jing, C., Shang, C., Yu, D., Chen, Y., Zhi, J.: The impact of different AR-HUD virtual warning interfaces on the takeover performance and visual characteristics of autonomous vehicles. Traffic Inj. Prev. **23**(5), 277–282 (2022). https://doi.org/10.1080/15389588.2022.2055752
3. Medenica, Z., Kun, A., Paek, T., Palinko, O.: Augmented reality vs. street views: a driving simulator study comparing two emerging navigation aids. In: Proceedings of the 13th International Conference on Human Computer Interaction with Mobile Devices and Services, pp. 265–274 (2011). https://doi.org/10.1145/2037373.2037414
4. Charissis, V., Papanastasiou, S., Vlachos, G.: Comparative Study of Prototype Automotive HUD vs. HDD: Collision Avoidance Simulation and Results (2008). https://doi.org/10.4271/2008-01-0203
5. Chauvin, C., Said, F., Langlois, S.: Augmented reality HUD vs. conventional HUD to perform a navigation task in a complex driving situation. Cogn. Technol. Work **25**(2–3), 217–232 (2023). https://doi.org/10.1007/s10111-023-00725-7
6. Smith, M., Streeter, J., Burnett, G., Gabbard, J.: Visual search tasks: the effects of head-up displays on driving and task performance. In: Proceedings of the 7th International Conference on Automotive User Interfaces and Interactive Vehicular Applications, pp. 80–87 (2015). https://doi.org/10.1145/2799250.2799291
7. Zhu, Y., Jing, Y., Jiang, M., Zhang, Z., Wang, D., Liu, W.: A experimental study of the cognitive load of in-vehicle multiscreen connected HUD. In: Soares, M.M., Rosenzweig, E., Marcus, A. (eds.) HCII 2021. LNCS, vol. 12781, pp. 268–285. Springer, Cham (2021). https://doi.org/10.1007/978-3-030-78227-6_20
8. Cheng, Y., Zhong, X., & Tian, L.: Does the AR-HUD system affect driving behaviour? An eye-tracking experiment study. Transp. Res. Interdisc. Perspect. **18**, 100767 (2023). https://doi.org/10.1016/j.trip.2023.100767
9. Hwang, Y., Park, B.-J., Kim, K.-H.: Effects of augmented-reality head-up display system use on risk perception and psychological changes of drivers. ETRI J. **38**(4) (2016). https://doi.org/10.4218/etrij.16.0115.0770
10. Park, J., Im, Y.: Visual enhancements for the driver's information search on automotive head-up display. Int. J. Hum.-Comput. Interact. **37**(18), 1737–1748 (2021). https://doi.org/10.1080/10447318.2021.1908667
11. Jose, R., Lee, G., Billinghurst, M.: A comparative study of simulated augmented reality displays for vehicle navigation. In: Proceedings of the 28th Australian Conference on Computer-Human Interaction, pp. 40–48 (2016). https://doi.org/10.1145/3010915.3010918
12. Young, J.Q., Van Merrienboer, J., Durning, S., Ten Cate, O.: Cognitive load theory: implications for medical education: AMEE Guide No. 86. Med. Teach. **36**(5), 371–384 (2014). https://doi.org/10.3109/0142159X.2014.889290
13. Xia, T., Lin, X., Sun, Y., Liu, T.: An empirical study of the factors influencing users' intention to use automotive AR-HUD. Sustainability (Basel, Switzerland) **15**(6), 5028 (2023). https://doi.org/10.3390/su15065028

DriveSense: A Multi-modal Emotion Recognition and Regulation System for a Car Driver

Lei Zhu[1], Zhinan Zhong[1], Wan Dai[1], Yunfei Chen[1], Yan Zhang[1](✉), and Mo Chen[2]

[1] School of Mechanical Engineering, Southeast University, Nanjing 211189, People's Republic of China
zhangyaner@seu.edu.cn
[2] College of Art and Design, Nanjing Tech University, Nanjing 210096, People's Republic of China

Abstract. Negative emotions significantly impact cognitive behavior and are a critical factor in driver safety and road traffic security. As intelligent driving systems evolve, the recognition and management of driver emotions has emerged as a crucial focus in automotive Human-Machine Interaction (HMI). We introduce DriveSense, an innovative emotion recognition and regulation system. DriveSense utilizes multi-modal data from onboard sensors, processes this data via deep learning in the cloud, and communicates with the HMI to implement regulation strategies. Our multi-modal emotion recognition model combines facial expression analysis and speech processing through Mel-frequency cepstral coefficients (MFCC), achieving a 60.37% accuracy on the RAVDESS dataset. We further validate DriveSense's utility through an experiment with 40 participants using a simulated driving scenario to test an adaptive music-based emotion regulation strategy. The results indicate that adaptive music can mitigate negative emotions effectively, underscoring DriveSense's potential to improve driver safety and secure driving practices.

Keywords: Driver emotion · Emotion recognition · Emotion regulation · Human-machine interaction · Deep learning

1 Introduction

With the continuous development of the economy and high technology, the automobile industry has witnessed a steady rise in terms of production and consumption. According to the data from the International Organization of Motor Vehicle Manufacturers, the global production of automobiles reached a total of 85.02 million units in 2022 [18], showing a year-on-year growth trend. Automobile s have become an essential part of people's daily life. However, the increasing number of traffic accidents poses a serious threat to the safety of people's lives and property.

H. Krömker (Ed.): HCII 2024, LNCS 14732, pp. 82–97, 2024.
https://doi.org/10.1007/978-3-031-60477-5_7

The World Health Organization's Global Status Report on Road Safety reveals that about 1.35 million people lose their lives in traffic accidents every year. It is critical to acknowledge that, alongside fatigue and distraction, a driver's compromised emotional state can greatly hinder their alertness and capacity for safe driving [10]. Consequently, this poses serious risks to both the driver's safety and the overall safety of road traffic. One of the main reasons is that the driving process involves a multitude of complex cognitive tasks, including but not limited to steering control, speed regulation, and road observation. Even minor disturbances caused by operational errors can result in catastrophic consequences, as emotions play a significant role in determining cognitive behaviors [15].

As a result, the recognition and intervention of driver emotions have become a prominent area of interest in automotive Human-Machine Interaction (HMI). Presently, automobile manufacturers are integrating features designed to enrich the driver's emotional experience within the cabin of the vehicle. For instance, NIO's voice assistant NOMI enhances the driver's emotional experience through the adaptive fun interfaces and voice interactions. Audi's Fit Driver system monitors real-time physiological data such as heartbeat and temperature to determine the driver's psychological pressure level, subsequently making timely adjustments to the car's temperature and even providing respiratory training videos to reduce safety hazards caused by negative emotions. Furthermore, some automakers are experimenting with concept cars. In 2020, Mercedes-Benz launched the Vision AVTR, a concept car engineered to detect a driver's emotional state by analyzing their breathing patterns. The car's adaptive lighting system reacts accordingly to these emotions. Advancing to 2023, BMW is poised to introduce the i Vision Dee, which promises to take vehicular technology to new heights. This vehicle promises to facilitate a more intuitive interaction between driver and machine by interpreting spoken words and facial expressions, effectively becoming a sentient companion for the driver's journey. These advancements underscore a transformative trend in automotive design, where cars are evolving from mere modes of transport into intelligent, emotionally aware entities. By comprehending user intentions and proactively responding, this transformation promotes driving safety and augments the overall user experience. Moreover, those advancements will expand the realm of interaction between humans and automobiles, fostering a more organic and meaningful human-car relationship [13].

A study [14] revealed that human emotions are expressed mainly through facial expressions (55%), voice (38%), and language (7%) in daily human communication. Visual and auditory clues are widely recognized as the most natural and meaningful channels for expressing emotions during communication [6]. Consequently, the field of emotion recognition primarily concentrates on two 'non-contact' methodologies: Visual Emotion Recognition (VER) [3] and Speech Emotion Recognition (SER) [5]. Moreover, certain studies employ 'contact' approaches, referred to as Physiological Emotion Recognition (PER) [27], which involve the collection and analysis of driving-related physiological signals using sensors. However, the practicality of this method is limited due to its interference with the driver's regular driving behavior, resulting in relatively lower accuracy compared to the 'non-contact' methods.

Nowadays, there has been a widespread adoption of deep learning-based methods in the field of emotion recognition. This enables the development of the construction of end-to-end neural network architectures that directly predict emotions without the requirement of complex feature engineering and selection processes. These approaches primarily use well-known pre-trained Convolutional Neural Networks (CNNs), such as VGG [22], ResNet [8] and GoogleNet [23]. The primary challenge lies in selecting a suitable loss function to address over-fitting concerns, particularly when working with relatively small datasets. In addition, Recurrent Neural Networks (RNN), including its variant known as Long Short Term Memory (LSTM) [9], have demonstrated notable performance in recognizing emotions from dynamic sequences or videos by establishing temporal correlations across the network's hidden layers. In the realm of visual emotion recognition, Kim et al. [11] proposed an end-to-end framework that integrates spatial features extracted by CNNs into LSTM layers, thereby encoding temporal characteristics. For speech emotion recognition, Trigeorgis et al. [24] developed an end-to-end SER system consisting of CNNs and LSTMs to automatically learn the optimal feature representation of speech signals directly from the original temporal representation.

Due to the innate human tendency to communicate and express emotional states through various modalities, there has been a growing research interest in multi-modal fusion methods for emotion analysis [25]. The appropriate selection of single-modal emotion input and the effective fusion strategies for multi-modal emotions are essential elements in the field of multi-modal-based emotion analysis, which often surpass the performance of single-modal emotion recognition systems. There are two main types of fusion strategies based on visual and auditory modalities. The first one is feature-level fusion, involving the combination of features extracted from multi-modal inputs into a unified feature vector, which is then fed into a classifier. For instance, Tzirakis et al. [24] employed CNN and ResNet to extract audio and video features correspondingly, connecting these two types of features to a 2-layer LSTM for predicting emotions. The other one is decision-level fusion, wherein the features of each modality are independently classified and subsequently fused into decision vectors to obtain the final result. Hao et al. [7] developed 4 sub-models utilizing SVMs and neural networks to capture manual-based and deep-learning-based audio-visual features. These sub-models were then fused through a hybrid integration algorithm to predict the ultimate emotion.

Upon detecting the driver's negative emotion, the car must quickly make adjustments and intervene in the driver's behavior to reduce the risk of accidents. Presently, the predominant methods for emotion regulation primarily center around sensory regulation, encompassing visual, auditory, olfactory, and tactile regulation. The visual modality serves as the primary channel for emotion perception in emotion monitoring systems. An efficient visual human-computer interface can significantly contribute to the regulation of the driver's emotions

[26]. For instance, a simple warning message displayed through the HMI can effectively inform the driver of the need for rest [1] to avoid fatigue while driving. Paredes et al. [19] implemented visual emotion regulation in a fully automated vehicle by employing a Virtual Reality (VR) headset. Compared to visual regulation, auditory regulation methods have the advantage of being spatially unrestricted and capable of receiving information from any direction. Nass et al. [17] found that a voice assistant that matches the driver's emotion has a positive effect on driving behavior and human-vehicle interaction. Odors possess the ability to stimulate the nervous system, making olfactory conditioning a crucial form of emotion regulation. Mustafa et al. [16] demonstrated that scents like vanilla and lavender can induce positive and relaxing feelings to some extent. Raudenbush et al. [20] found that peppermint and cinnamon scents aid in reducing frustration, enhancing driver focus on the driving task, and alleviating feelings of nervousness and anxiety. Additionally, the effectiveness of tactile modulation lies in the presence of various receptors in the skin. Schmidt et al. [21], in their study on temperature modulation, observed that cool airflow induced arousal and promoted better driving behavior.

In this study, we introduce DriveSense, a cutting-edge system designed for intelligent emotion recognition and regulation within vehicles. DriveSense aims to bolster driving safety by capturing the driver's emotional state in real time via onboard sensors, then analyzing the data using deep learning algorithms in the cloud. The system is interfaced with the vehicle's HMI to provide appropriate emotional regulation strategies. Our work presents a practical approach to applying emotion recognition and regulation technology in a vehicular setting. We have developed a multi-modal emotion recognition model that synthesizes facial expression and speech data. The system's effectiveness is validated through a series of experiments focused on emotion regulation.

2 The DriveSense Framework

2.1 System Overview

Illustrated in Fig. 1, the architecture of the DriveSense emotion recognition and regulation system comprises four layers: the Driver layer, the Cloud layer, the HMI layer, and the In-cabin Sensor layer.

The Driver layer is tailored for operators of L2-L3 level intelligent vehicles, where the driver is the main vehicle controller and their behavior is vital for driving safety. DriveSense activates in response to the driver's negative emotional states, providing timely intervention to maintain safety.

The Cloud layer serves as the core module of the DriveSense system, providing an emotion detection capability based on in-cabin sensors. The contemporary approach to driver emotion involves the application of deep learning-based techniques, utilizing an end-to-end neural network to predict the ultimate emotion. This process necessitates substantial computational power to support sensing, processing, and fusion of data. Therefore, the system employs the emotion detection module situated in the cloud layer, enabling real-time collection of driver

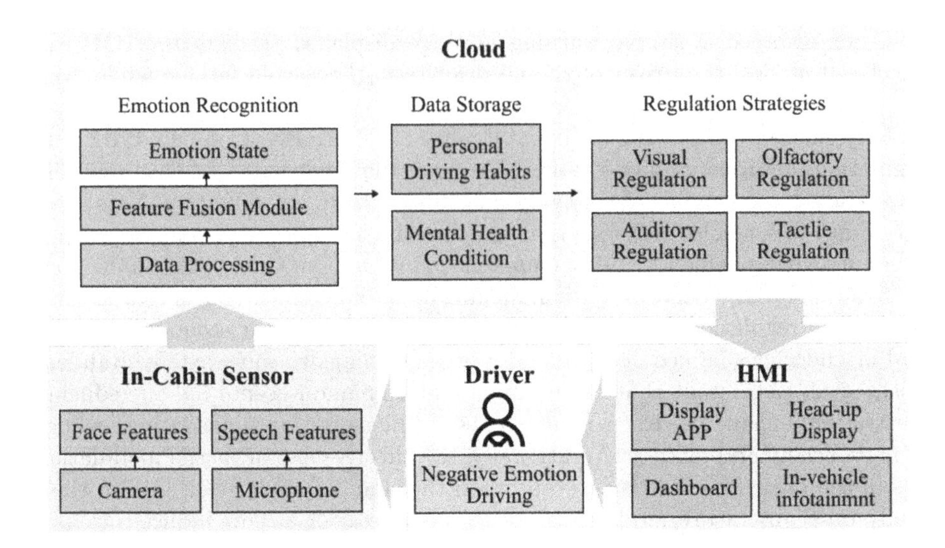

Fig. 1. The architecture of the DriveSense system.

emotion data from multiple sensors and generating emotion recognition outcomes through a well-designed emotion recognition model.

The HMI layer serves as the intermediary between the driver and the overall system. Particularly in cases where the vehicle's automation is not yet fully realized, the design and functionality of the HMI are critical to ensure that the driver and the vehicle-based automation system collaborate safely. To achieve this, the system integrates various interactive visual interfaces, as illustrated in Fig. 2, including the head-up display, dashboard, display APP, and in-vehicle infotainment. Through multi-modal interaction, these interfaces actively monitor the driver's emotions and employ strategies such as adaptive music, ambient light, and UI intervention to regulate the driver's emotional state effectively, thereby ensuring emotional stability during driving.

The In-cabin Sensor layer is the basis for realizing DriveSense functions, including physical devices and technical support. Specifically, it involves equipment for acquiring speech and facial features. It also includes network communication technology, multimedia technology, and other aspects of support. Furthermore, the security and stability of the system should be taken into account to ensure its reliable operation.

2.2 Emotion Recognition Model

Emotion definition is the key to establishing a standard for emotion recognition. We adopt the discrete emotion model proposed by Ekman [4] to quantitatively describe emotions, including eight basic emotional states: anger, disgust, fear, happiness, sadness, surprise, calm, plus neutral emotions. In the DriveSense system, we utilize sensors to extract emotional information from the driver's facial

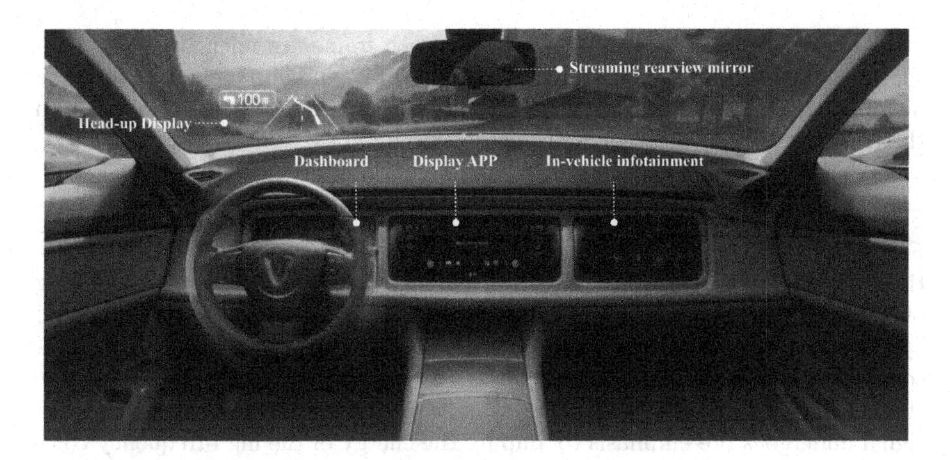

Fig. 2. HMI interaction model.

expressions and speech. The overall framework of the emotion recognition model is shown in Fig. 3.

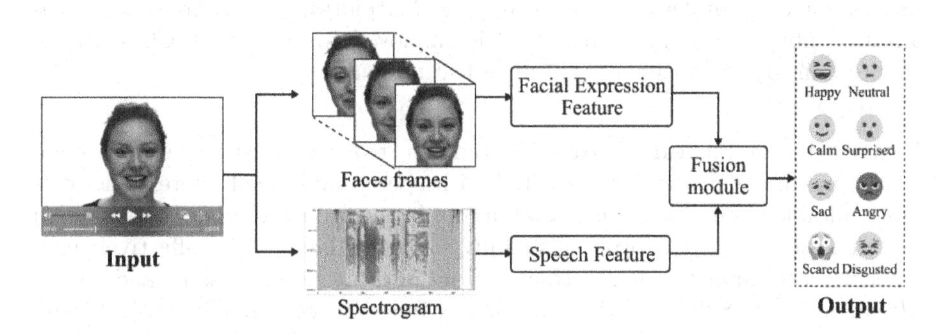

Fig. 3. The framework of the emotion recognition model.

Facial Expression Feature Extraction. The process of extracting facial expression features involves two steps. In the first step, facial expression features are extracted from a single-frame image obtained from the input video. This involves cropping the original image to obtain a consistently sized region that encompasses only the face. To achieve this, we utilized ResNet [8], a well-established CNN feature extraction network known for its exceptional proficiency in feature extraction and facial expression recognition tasks. Specifically, we employed ResNet-18, which underwent pre-training on the extensive ImageNet dataset, comprising 1.4 million labeled images. The feature extraction module receives 74×74-pixel face images as input and produces a 512-dimensional 7×7-pixel feature map as output. In the second step, an LSTM network [9] is utilized to learn the comprehensive representation of the entire sequence of video frames after obtaining the feature maps for each frame. To mitigate overfitting, the Dropout regularization method is employed by randomly deactivating some neuron outputs during training. The resulting feature sequence is then fed into the fusion module for further processing.

Speech Feature Extraction. In the speech feature extraction section, Mel-Frequency Cepstral Coefficients (MFCC) are selected as the feature representation. This selection is motivated by the superior robustness and ease of processing offered by MFCC features, in comparison to other alternatives like chromatic and spectral features. To preprocess the speech signal, the librosa library, an open-source tool dedicated to audio analysis and processing, is employed. Specifically, the librosa library is utilized to read audio files. Initially, the original speech signal undergoes pre-emphasis to amplify the energy of the high-frequency component, thereby rectifying the high-frequency attenuation problem inherent in the transmission link. Subsequently, the speech signal is partitioned into frames of fixed length through framing techniques, with a window function applied. Each frame is then transformed into a frequency domain representation using the Short-time Fourier transform (STFT), enabling the calculation of the amplitude and power spectrum. Finally, the power spectrum is transformed into an image-format input feature map using the Matplotlib open-source data visualization library. Similarly, Resnet-18 is employed to extract MFCC features, resulting in 512-dimensional 7×7-pixel feature maps.

Feature-Level Fusion Module. Given the strong correlation between emotion representation vectors associated with speech and facial expressions, it is more suitable to employ feature-level fusion. In the proposed model, the emotion feature fusion module incorporates a fully connected layer, which effectively concatenates the input features extracted from both facial expressions and speech MFCC features. Subsequently, the softmax function was employed to calculate the probabilities corresponding to the eight discrete emotions. As shown in Eq. 1, the output array denoted as $\mathbf{z}$ encapsulates the probability scores assigned to each emotion predicted by the model. The probability allocated to the ith emotion class equals its specific score divided by the sum of scores across all emotions. This method is widely accepted for normalizing probabilities between 0 and 1.

$$\sigma(\mathbf{z})_i = \frac{e^{z_i}}{\sum_{n=1}^{K} e^{z_n}} \tag{1}$$

3 Evaluation of Emotion Recognition Model

This section is to evaluate the performance of the proposed multi-modal driver emotion recognition model. The experiments are carried out using a publicly accessible multimodal emotion dataset.

3.1 Experiment Environment and Data Set

In this experiment, the Google Colab platform is selected as the experimental environment. Google Colab is a cloud computing platform offered by Google, providing convenient access to cloud GPU and TPU resources. It enables users to write, execute, and share code within a browser-based interface. The specific experimental setup in this research employs a Tesla A100 GPU and PyTorch version 2.0.1.

The RAVDESS public dataset [12] (Fig. 4) was selected for training and validation of the model due to its accessibility and availability of raw data. This dataset comprises video recordings of 24 proficient actors, evenly distributed across genders (12 females and 12 males), delivering dialogues in various emotional states. Specifically, the dataset encompasses eight types of emotional tasks, namely calm, happy, sad, angry, fearful, surprised, disgusted, and neutral. Each actor contributed 60 video sequences.

Fig. 4. Examples of the RAVDESS dataset.

3.2 Evaluation Metrics

Taking the binary classification problem as an example, instances are classified as either positive or negative examples. By comparing the actual instances with the predicted results, we can categorize them into four types: True Positives (TP), True Negatives (TN), False Positives (FP), and False Negatives (FN), as illustrated in the confusion matrix presented in Table 1.

In deep learning-based classification tasks for emotion recognition, Accuracy and F1-score are commonly employed as evaluation metrics to assess the performance of models. Accuracy is the proportion of the correctly predicted result to the total number of samples, as shown in Eq. 2.

Table 1. Confusion matrix for classification results.

	Predicted results	
Actual instances	Positive	Negative
Positive	TP	FN
Negative	FP	TN

$$\text{Accuracy} = \frac{TP + TN}{TP + TN + FP + FN} \tag{2}$$

However, Accuracy alone may not be sufficient for evaluating a model's performance, especially when dealing with imbalanced datasets. Therefore, the F1-score, as a metric specifically designed to achieve an optimal trade-off between Recall and Precision, is also used. The F1-score considers both precision and recall, providing a comprehensive measure of the model's effectiveness. Precision (Eq. 3) refers to the proportion of true positive instances among all the instances classified as positive, while Recall (Eq. 4) denotes the proportion of true positive instances out of all the positive instances.

$$\text{Precision} = \frac{TP}{TP + FP} \tag{3}$$

$$\text{Recall} = \frac{TP}{TP + FN} \tag{4}$$

$$F1 = 2 \cdot \frac{Precision \cdot Recall}{Precision + Recall} \tag{5}$$

Furthermore, to enhance the accuracy of the model, we employed the cross-entropy loss function as the objective function during the training process. This loss function quantifies the disparity between the predicted results and the true labels, facilitating parameter updates that aim to minimize the loss value. To expedite the convergence of the model and achieve improved results, we utilized the Adam optimizer for the parameter updating procedure.

3.3 Analysis of Experimental Results

The key innovation of the emotion recognition model presented in this paper lies in its ability to fuse multiple emotional modal features, encompassing both facial expressions and speech features. To assess the individual contributions of each modality towards the overall emotion recognition task, we conducted a comprehensive performance analysis of the model under three different scenarios: 'FE Only' denotes the exclusive utilization of facial expression features for model training. 'S Only' signifies the sole incorporation of speech features into the training process. 'FE+S' represents the integration of multi-modal features, encompassing both facial expression and speech features. To evaluate the efficacy of the proposed approach, we employed Accuracy and F1-score metrics, and the results of these comparative experiments are exhibited in Table 2.

Table 2. Results of the emotion recognition model comparative experiment.

Model	Accuracy	F1 score
FE+S	0.6037	0.5479
S only	0.5403	0.4632
FE only	0.3882	0.3017

The model's performance achieved a rate of 54.03% when relying on speech features, which was notably 15.21% higher compared to using facial expressions features alone. However, upon combining bimodal features, the model's accuracy increased substantially to 60.37%. These findings underscore the complementary nature of speech and facial modalities in the emotion recognition tasks, further emphasizing the potential for feature fusion to enhance overall model performance. Moreover, it is crucial to acknowledge that multi-modal emotion recognition tasks often involve high-dimensional features, posing challenges related to computational and storage resources. Consequently, system developers must consider the limitations imposed by computational and storage resources during the development process.

4 Evaluation of an Adaptive Music-Based Emotion Regulation Strategy

In this section, we chose the auditory interface incorporating an adaptive music-based regulation strategy [2], combined with the emotion recognition model, to validate its effectiveness in regulating the driver's negative emotions.

4.1 Participants

40 participants (20 males and 20 females) from 20 to 27 years old, who possessed a valid driving license and had over 1 year of driving experience, were recruited to participate in this study. All participants were in good health, without any visual or hearing impairments. Before their involvement in the experiment, they were required to provide informed consent by signing a consent form and also furnish a registration form containing basic personal information.

4.2 Experiment Design

The participants were required to accomplish driving tasks under specific negative emotional states. To effectively induce the emotions of anger, fear, and disgust, a meticulous selection process was undertaken to identify three videos as the stimuli for emotion induction, as presented in Table 3. Moreover, adaptive music was chosen for regulating emotions. In order to minimize any potential interference arising from lyrical content, the experiment employs tranquil instrumental music "Boundless Horizon" as the material for alleviating negative emotions.

Table 3. A brief description of the selected driver's negative emotions video stimulus.

Target mood	Content of material	Duration/s
Angry	A woman abused a traffic police officer and her own country after she was stopped for violating traffic rules	78
Scared	Witnessing the scene of multiple tragic traffic accidents	60
Disgusted	The driver finds a passenger lying on the passenger seat with bare feet	88

This experiment took place within a simulated environment based on Horizon 5, as depicted in Fig. 5, consisting of a Xiaomi 50-in. monitor, a Tumast T248 driving simulator, and a Logitech camera.

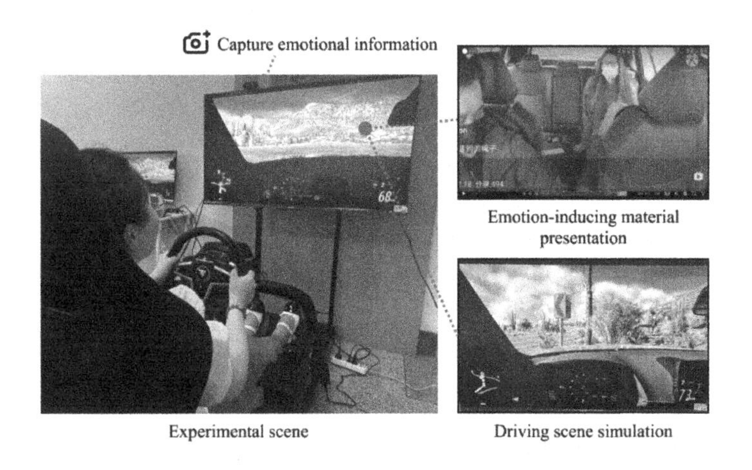

Fig. 5. Experiment setup for emotion regulation

The experimental procedure comprises the following steps:

Step 1: Participants familiarized themselves with the driving simulator by completing a 2 km driving task on the practice road, following the designated route in Horizon 5. After completing the exercise, they were given a 2-min break.

Step 2: Participants were instructed to carry out three rounds of experiments, each consisting of two groups as follows:

Group A (the control group): Participants watched the emotion-inducing video and then drove on the designated scenario road for 2 min, maintaining a speed of no more than 60 km/h. Upon completing the driving task, they were asked to recall their emotional state and provide a 30-s evaluation of their driving experience. During this period, emotion recognition was conducted based on the emotion recognition model.

Group B (the experimental group): Participants followed the same instructions as Group A. However, in contrast to Group A, adaptive music was applied during this period to effectively regulate the participants' emotions.

Step 3: These steps were repeated until all three rounds of the experimental task were completed for each emotional state. The experimenter recorded the designated emotional scores for each group.

Step 4: The data were statistically analyzed using SPSS software. The analysis aimed to examine the emotional changes experienced by the participants and explore the effectiveness of the adaptive music-based emotion regulation strategy.

4.3 Analysis of Experiment Results

Considering the strain on computational resources posed by excessively long videos, we divided the 30-s recordings, encompassing both speech and facial information from the 40 subjects, into equal-duration segments, as depicted in Fig. 6. Subsequently, these segments were input into the emotion recognition model to acquire emotion predictions. The resulting emotion prediction for the input video is derived by averaging the predictions obtained from multiple segments.

We conducted paired-samples *t*-test and Cohen's *d* analysis to assess the significance and effect size of the difference between Group A and Group B in terms of negative emotions. In the null hypothesis for *t*-test, it was assumed that no significant difference existed between the two groups. The *t*-value calculation as shown in Eq. 6 includes the means of paired-samples difference, standard deviation, and sample size. Typically, a significance level (α) of 0.05 is employed. If the calculated *t*-value is less than the critical value at the α level (indicating p >0.05), the null hypothesis is accepted. Conversely, if the *t*-value exceeds the critical value (indicating p <0.05), the null hypothesis is rejected, implying a significant difference between the means of Group A and Group B.

$$t = \frac{|\overline{x_1} - \overline{x_2}|}{\sqrt{\frac{s_1^2}{n_1} + \frac{s_2^2}{n_2}}} \tag{6}$$

Cohen's *d* analysis is employed to quantify the effect size of the mean difference between two groups, as depicted in Eq. 7. Here, $\overline{x_1}$ and $\overline{x_2}$ represent the means of the two groups, while s denotes the pooled standard deviation as calculated with Eq. 8. This approach addresses the issue of over-reliance on p-values and helps mitigate misinterpretation of findings. Conventionally, effect sizes such as small, medium, and large are defined by Cohen's *d* values of 0.2, 0.5, and 0.8, respectively. In terms of the difference between the observed values, medium

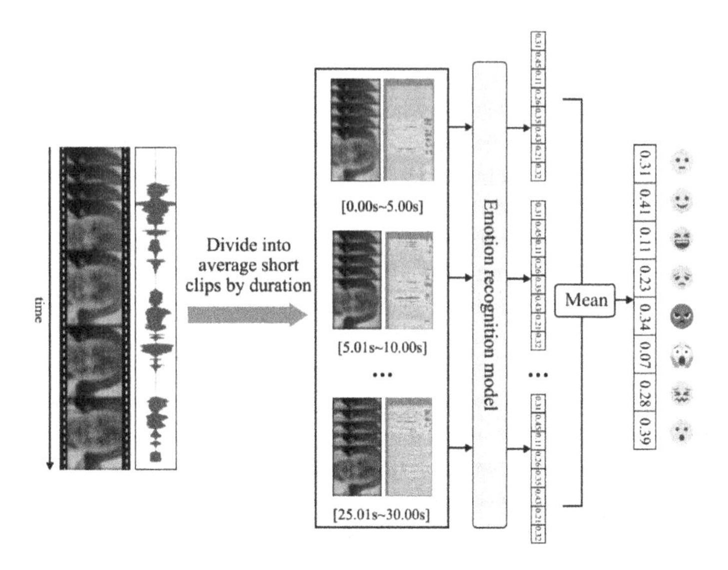

Fig. 6. Emotion scores prediction process.

and large effect size levels suggest stronger reliability compared to small effect size values.

$$d = \frac{\overline{x_1} - \overline{x_2}}{s} \tag{7}$$

$$s = \frac{\sqrt{(n_1 - 1)\, s_1^2 + (n_2 - 1)\, s_2^2}}{n_1 + n_2 - 2} \tag{8}$$

Based on the findings presented in Table 4 and Fig. 7, the experimental data analysis reveals statistically significant results for anger, fear, and disgust emotions in Groups A and B, with p-values of 0.000 (<0.050), 0.001 (<0.050), and 0.005 (<0.050), respectively. These results clearly demonstrate a significant difference in emotional states between environments with and without adaptive music. In addition, Cohen's d value for anger surpasses that of the other two emotions, indicating that adaptive music exhibits superior effectiveness in regulating anger. In conclusion, the introduction of adaptive music successfully alleviates negative emotions among drivers, particularly in terms of anger regulation, as substantiated by the aforementioned statistical analysis.

Moreover, it is crucial to recognize that music requires a certain amount of cognitive resources. Given the limited capacity of cognitive resources, the introduction of adaptive music into the driving environment has the potential to disrupt the ongoing driving task and create a safety hazard. While the results of this experiment suggest a positive impact of adaptive music on drivers' emotional well-being, it is important to recognize that there may also be adverse consequences, such as interference with the cognitive resources dedicated to driving. These factors underscore the necessity for a comprehensive evaluation of the

Table 4. p-value, Cohen's d value, and effect size between Group A and Group B

	Angry		Scared		Disgusted	
	Group A	Group B	Group A	Group B	Group A	Group B
Mean	0.598	0.365	0.379	0.298	0.441	0.370
Std. deviation	0.036	0.039	0.044	0.038	0.041	0.039
p**-value**	0.01×10^{-3}		1.37×10^{-3}		4.59×10^{-3}	
Cohen's d **value**	6.21		1.97		1.77	
Effect size	Large		Large		Large	

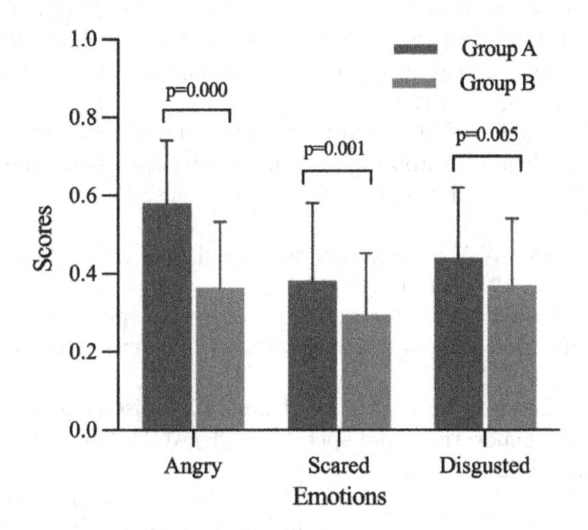

Fig. 7. Statistical analysis of emotion scores.

overall implications when considering the suitability of incorporating adaptive music in the driving context.

5 Conclusion

This paper presents a framework for the DriveSense in-vehicle intelligent emotion recognition and regulation system, along with a series of studies conducted to validate its feasibility. Our approach proposes a multi-modal fusion method for emotion recognition that integrates facial expression and speech features. The findings demonstrate that integrating these two independent modalities into a multi-modal approach enhances the accuracy of emotion recognition and yields more reliable results. Furthermore, concerning the design parameters of the auditory interface, it was discovered that adaptive music plays a constructive role in regulating drivers' negative emotions to a certain extent. As such, DriveSense can be regarded as an effective normative tool for promoting safe driving among drivers. While this study concentrated on fundamental discrete emotions, future research should explore expanding the system's applicability to a wider spectrum of emotions. Additionally, there is room for innovation in the realm of emotion regulation interaction, especially concerning in-vehicle HMI.

Disclosure of Interests. The authors have no competing interests to declare that are relevant to the content of this article.

References

1. Braun, M., Schubert, J., Pfleging, B., Alt, F.: Improving driver emotions with affective strategies. Multimodal Technol. Interact. **3**(1), 21 (2019)
2. Braun, M., Weber, F., Alt, F.: Affective automotive user interfaces-reviewing the state of driver affect research and emotion regulation in the car. ACM Comput. Surv. (CSUR) **54**(7), 1–26 (2021)
3. Corneanu, C.A., Simón, M.O., Cohn, J.F., Guerrero, S.E.: Survey on RGB, 3D, thermal, and multimodal approaches for facial expression recognition: history, trends, and affect-related applications. IEEE Trans. Pattern Anal. Mach. Intell. **38**(8), 1548–1568 (2016)
4. Ekman, P., Friesen, W.V.: Constants across cultures in the face and emotion. J. Pers. Soc. Psychol. **17**(2), 124 (1971)
5. El Ayadi, M., Kamel, M.S., Karray, F.: Survey on speech emotion recognition: features, classification schemes, and databases. Pattern Recogn. **44**(3), 572–587 (2011)
6. Eskimez, S.E., Maddox, R.K., Xu, C., Duan, Z.: Noise-resilient training method for face landmark generation from speech. IEEE/ACM Trans. Audio Speech Lang. Process. **28**, 27–38 (2019)
7. Hao, M., Cao, W.H., Liu, Z.T., Wu, M., Xiao, P.: Visual-audio emotion recognition based on multi-task and ensemble learning with multiple features. Neurocomputing **391**, 42–51 (2020)
8. He, K., Zhang, X., Ren, S., Sun, J.: Deep residual learning for image recognition. In: Proceedings of the IEEE Conference on Computer Vision and Pattern Recognition, pp. 770–778 (2016)
9. Hochreiter, S., Schmidhuber, J.: Long short-term memory. Neural Comput. **9**(8), 1735–1780 (1997)
10. Hu, T.Y., Xie, X., Li, J.: Negative or positive? The effect of emotion and mood on risky driving. Transport. Res. F: Traffic Psychol. Behav. **16**, 29–40 (2013)
11. Kim, D.H., Baddar, W.J., Ro, Y.M.: Micro-expression recognition with expression-state constrained spatio-temporal feature representations. In: Proceedings of the 24th ACM international conference on Multimedia, pp. 382–386 (2016)
12. Livingstone, S.R., Russo, F.A.: The Ryerson audio-visual database of emotional speech and song (Ravdess): a dynamic, multimodal set of facial and vocal expressions in north American English. PLoS ONE **13**(5), e0196391 (2018)
13. McDuff, D., Czerwinski, M.: Designing emotionally sentient agents. Commun. ACM **61**(12), 74–83 (2018)
14. Mehrabian, A.: Communication without words. In: Communication Theory, pp. 193–200. Routledge (2017)
15. Mesken, J., Hagenzieker, M.P., Rothengatter, T., De Waard, D.: Frequency, determinants, and consequences of different drivers' emotions: an on-the-road study using self-reports,(observed) behaviour, and physiology. Transport. Res. F: Traffic Psychol. Behav. **10**(6), 458–475 (2007)

16. Mustafa, M., Rustam, N., Siran, R.: The impact of vehicle fragrance on driving performance: what do we know? Procedia. Soc. Behav. Sci. **222**, 807–815 (2016)
17. Nass, C., et al.: Improving automotive safety by pairing driver emotion and car voice emotion. In: CHI 2005 Extended Abstracts on Human Factors in Computing Systems, pp. 1973–1976 (2005)
18. OICA: Oica correspondents survey: World motor vehicle production by country/region and type (2022). https://www.oica.net/category/production-statistics/2022-statistics/
19. Paredes, P.E., et al.: Driving with the fishes: towards calming and mindful virtual reality experiences for the car. In: Proceedings of the ACM on Interactive, Mobile, Wearable and Ubiquitous Technologies, vol. 2, no. 4, pp. 1–21 (2018)
20. Raudenbush, B., Grayhem, R., Sears, T., Wilson, I.: Effects of peppermint and cinnamon odor administration on simulated driving alertness, mood and workload. N. Am. J. Psychol. **11**(2), 245–245 (2009)
21. Schmidt, E., Decke, R., Rasshofer, R., Bullinger, A.C.: Psychophysiological responses to short-term cooling during a simulated monotonous driving task. Appl. Ergon. **62**, 9–18 (2017)
22. Simonyan, K., Zisserman, A.: Very deep convolutional networks for large-scale image recognition. arXiv preprint arXiv:1409.1556 (2014)
23. Szegedy, C., et al.: Going deeper with convolutions. In: Proceedings of the IEEE Conference on Computer Vision and Pattern Recognition, pp. 1–9 (2015)
24. Tzirakis, P., Trigeorgis, G., Nicolaou, M.A., Schuller, B.W., Zafeiriou, S.: End-to-end multimodal emotion recognition using deep neural networks. IEEE J. Sel. Top. Signal Process. **11**(8), 1301–1309 (2017)
25. Tzirakis, P., Zhang, J., Schuller, B.W.: End-to-end speech emotion recognition using deep neural networks. In: 2018 IEEE International Conference on Acoustics, Speech and Signal Processing (ICASSP), pp. 5089–5093. IEEE (2018)
26. Wang, Y., et al.: A systematic review on affective computing: emotion models, databases, and recent advances. Inform. Fusion **83**, 19–52 (2022)
27. Zhang, J., Yin, Z., Chen, P., Nichele, S.: Emotion recognition using multi-modal data and machine learning techniques: a tutorial and review. Inform. Fusion **59**, 103–126 (2020)

Human Factors in Automated Vehicles

A Two-Loop Coupled Interaction System Design for Autonomous Driving Scenarios

Mingyu Cui[1], Yahui Zhang[2], and Kejia Zhang[1(✉)]

[1] The Future Laboratory, Tsinghua University, Beijing, China
354083048@qq.com
[2] Tsinghua University, Beijing, China

Abstract. The development of automated driving technology gradually detaches the driver from the driving task and allows more interaction in the intelligent cockpit. Problems ensue, as redundancy of interaction information, enhanced cognitive load, and imbalance of resource allocation become increasingly prominent, reducing driver attention and operational efficiency and leading to an imbalance between interactivity and safety. The automotive industry urgently needs to redefine the human-vehicle interaction relationship. In this paper, we construct an interaction design set concept, sort out the interaction-related factors of autonomous driving scenarios, and establish a human-centered intelligent cockpit ring interaction set and a vehicle-centered autonomous driving ring interaction set. The relationship between driving tasks, cognitive contention, user experience and human-machine ergonomics is explored. Visualization through the sets redefines the design methodology and evaluation methodology of interaction systems for present and future autonomous driving scenarios, providing guidance to designers.

Keywords: Autonomous driving · Interactive Systems · Design Methods · Two-loop

1 Introduction

In recent years, the connected communication digital technology represented by artificial intelligence, big data, blockchain, cloud computing, Internet of Things and other technologies has developed rapidly [1]. The rapid development of information technology has given rise to a variety of functions such as automobile automatic driving, intelligent network connection, intelligent cockpit interaction, etc., and vehicles are gradually moving from mechanization and intelligence to informationlization.

1.1 Development of Automobile Automatic Driving Technology

The level of automatic driving technology is usually divided according to the SAE international standard (Society of Automotive Engineers), which defines 6 different levels from level 0 to level 5. Level L0 - full manual driving, level L1 - assisted driving, level L2 - assisted driving, level L3 - assisted driving, level L4 - assisted driving, level L5

H. Krömker (Ed.): HCII 2024, LNCS 14732, pp. 101–115, 2024.
https://doi.org/10.1007/978-3-031-60477-5_8

- assisted driving, level L6 - assisted driving, and level L7 - assisted driving. L0 - fully manual driving, L1 - assisted driving, L2 - partially automated driving, L3 - conditional automated driving, L4 - highly automated driving, and L5 - fully automated driving. The basic situation is shown in the Table 1.

Table 1. Summary of Autonomous Driving Scenarios.

Level	Basic information	Technical realization	Human-machine relationship
L0	Fully manual driving		Manual driving
L1	The driving system is able to perform 1 of the steering and acceleration/deceleration operations based on the environmental information, while the other operations are still performed by the driver	Anti-lock Braking System (ABS), Electronic Stability System (ESP)	Autonomous Driver Assistance
L2	The driving system is able to perform steering and acceleration/deceleration operations based on environmental information, while other operations are still performed by the driver	Adaptive Cruise Control, Active Lane Keeping System, Automatic Brake Assist, and Auto Parking System	Internet-connected driver assistance
L3	The driving system is capable of performing all driving operations, and the driver is required to provide appropriate takeover as requested by the system	Fully automated high-speed navigation pilot (NGP, Navigation Guided Pilot) + fully automated valet parking + automated driving assistance on congested urban roads (TJP)	Man-machine co-driving
L4	The driving system performs all driving maneuvers, and in certain environments the system will request a response from the driver	**Sensing Layer:** High Precision Sensors: LIDAR, Millimeter Wave Radar, Camera, Ultrasonic Radar, GNSS/IMU **Decision-making layer:** Navigation and positioning, high-precision maps + sensors, central processor + algorithms **Execution layer:** Electronic drive, electronic brake, electronic steering [2]	
L5	Driving system for all road conditions		Intelligent Body Autonomous Driving

Four forms of human-machine relationships are formed in the six autonomous driving tiers of L0-L5: autonomous driver assistance, networked driver assistance, human-machine co-driving, and highly automated/driver less. Autonomous Driver Assistance System S (ADAS) is a system that utilizes in-vehicle sensors for environment perception and provides driving operation assistance to the driver. Internet-connected driver assistance system refers to the use of information and communication technology (Information Communication Technology, ICT) for real-time perception of the environment around the vehicle, and predict the future movement of the surrounding vehicles, so as to provide driving operation assistance to the driver's system. [3] Human-machine co-driving technology aims to enable intelligent vehicles to realize intelligent driving, intelligent systems and drivers can work together to complete the driving task.

1.2 Development of Automobile Automatic Driving Technology

Human-Computer Interaction is a study that focuses on the interaction patterns between humans and machines. Automotive interaction design is the study of the relationship between human and automobile in driving scenarios. At present, some major areas have been developed, such as automated human-machine interaction (aHMI), vehicle human-machine interaction (vHMI), infotainment interaction (iHMI), external human-vehicle interaction (eHMI), and dynamic human-machine interaction (dHMI) [4].

In Intelligent Cockpit Interaction. Automated Human-Machine Interaction (aHMI) is an interactive display about the automated driving system of the vehicle, which shows the automated driving status of the car, transmits the driving information of the car, and at the same time monitors the status of the driver in order to provide real-time feedback from the vehicle, and completes the conversion of the driving roles of the driver and the vehicle in real time. Vehicle Human-Machine Interaction (vHMI) transmits automobile information, such as road conditions, air conditioning, fuel, and vehicle speed, through single-channel or multi-channel interactions, such as visual, auditory, and tactile. Infotainment interaction (iHMI) is more inclined with multi-channel, intelligent, immersive human-vehicle interaction design solutions, where interaction designers add personalized functions such as work-study, leisure and entertainment in the intelligent cockpit.

In Automotive External Interaction. External human-vehicle interaction (eHMI) includes the interaction between vehicles and vehicles, pedestrians, and the environment, and the communication between vehicles and vehicles, pedestrians, and the environment is accomplished through the light display, external projection, and other interaction external to the vehicle, which is conducive to reducing the conflict of road use. For example, self-driving vehicles increase traffic flow and safety by negotiating the right-of-way in narrow roadways by showing their intention to yield through eHMI. Dynamic human-machine interaction (dHMI) is a vehicle dynamics-following, functional human-machine interface for communicating vehicle dynamics information to road users. The development of autonomous driving technology is gradually removing the driver from the driving task and allowing for more interaction in the intelligent cockpit, while the vehicle is taking on more driving tasks and is also taking on more interactive information transfer.

1.3 Needs

The development of autonomous driving technology allows drivers to gradually detach themselves from the driving task and can interact more in the intelligent cockpit, while the vehicle is taking on more driving tasks and is also taking on more interactive information transfer.

The automobile in the mechanical era is a one-way linear interaction from human to vehicle, and in the electronic era, it is expanded to a two-way feedback interaction form which is mainly linear. In the intelligent era, researchers began to focus on human entertainment, socialization, and guidance to build a multi-linear interaction between people and vehicles. With the development of autonomous driving and intelligent network connection, the driver is gradually detached from the driving task and can interact more in the intelligent cockpit, while the vehicle is taking on more driving tasks and is also taking on more interactive information transfer. The original linear interaction system has gradually become large and complex, can such an interaction design method still accurately represent automotive interaction relationships? At this stage, there are more metrics in the form of interaction, such as user experience, human-computer performance, cognitive psychology, and trust mechanisms, how can such information be presented in a linear design, as Fig. 1

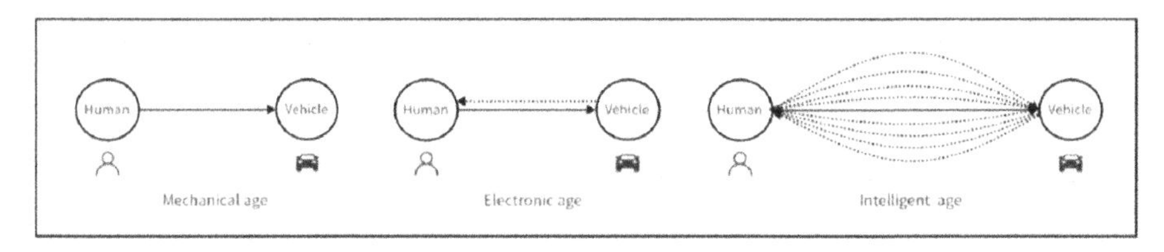

Fig. 1. A linear human-vehicle interaction design methodology

Driving a car is a complex operational behavior that requires a high degree of concentration. Are automotive companies' attempts to add interactive features to the cockpit affecting the driver's concentration and operational efficiency, leading to an imbalance between interactivity and safety. Are designers' pie-in-the-sky visions of smart cockpits in real-world applications compatible with the development of autonomous driving technology?

The original linear design of automobiles does not seem to be able to accurately describe these relationships and provide designers with effective design measurements. In the era of automotive informatization, it is necessary for us to step out of the original framework and redefine and rethink the forms of interaction among people, vehicles, and environments.

2 Related Research

2.1 Automotive Interaction System Design Methodology

Yang et al. [5] Human-machine interaction is the main channel for the expression of the performance and system characteristics of the whole ICV system, and all the interaction tasks are concentrated in the HMI, which is evaluated by establishing a trust model for the HMI through a linear approach. TAN [6] establishes a system model for the design of automotive voice interactions through a linear and frame approach; Schmidt [7] argues that the iHuman-Machine-Interface system may include the vehicle's external interactions with the users around it. Expressing external HMIs through linear system diagrams for road safety and efficiency in future highly automated road scenarios. The effect of the linear interaction system is shown in the Fig. 2.

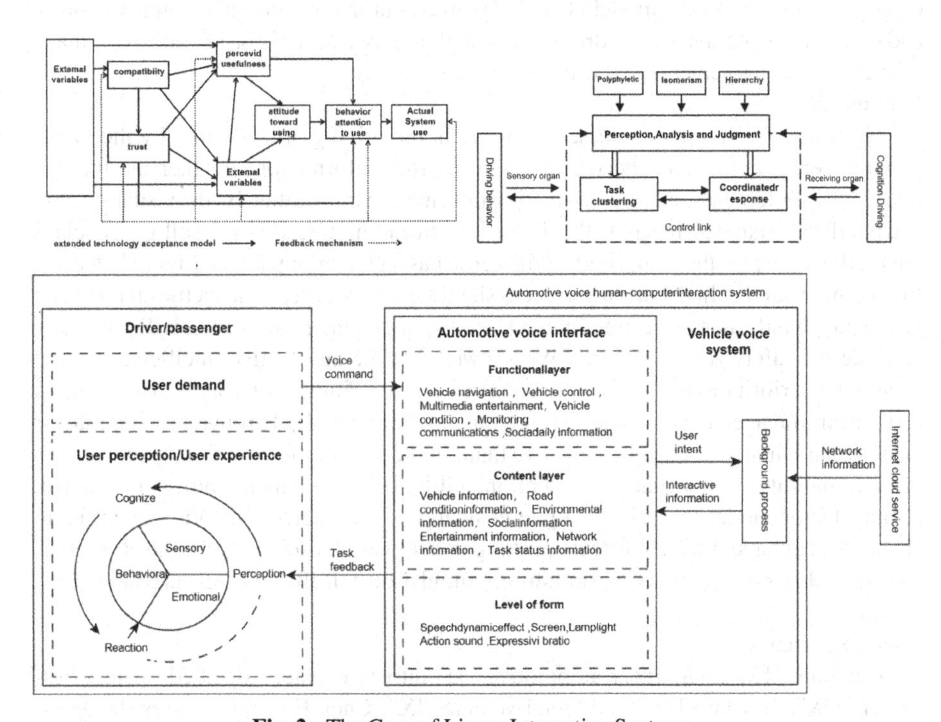

Fig. 2. The Case of Linear Interactive Systems

2.2 Research on Automotive Interaction Effects

Cognitive Psychology

Cognitive factors play an important role in the human-computer interaction design of intelligent networked vehicles. Designers need to fully consider users' needs and psychological characteristics from the three aspects of cognitive load attention allocation and automation trust in order to provide a better interaction experience. Cognitive competition reflects the contradiction between rich information dissemination and limited human information processing ability [8]. Cognitive competition in the field of automotive interaction is manifested in the behavior of interactive information competing for the limited cognitive resources of the driver.

FAAS [9] in their study pointed out that other non-driving human-computer interaction behaviors can distract the driver's attention to a certain extent when the driver is focusing on the driving task, which can cause safety hazards. Dukes suggested that the driver's cognitive, emotional and physiological perceptions in the process of intelligent driving affect the driver's take-over driving behaviors to a certain extent. Zhao [10], through the comparative analysis of theoretical models of driving behaviour, also pointed out that the driver's cognitive, emotional and physiological perceptions in the process of intelligent driving would affect their takeover driving behaviour to a certain extent. Chen's theoretical model [11] also pointed out that the driver's cognitive ability and emotional state during the driving task will largely affect his or her ability to make correct driving judgements and driving decisions.

Ergonomics

The contact medium of interaction driver in the driving process, first by the visual system to observe the road ahead and around the road information, and then use the arm muscles to implement the decision-making information transmitted by the visual system, and finally to make steering and other behaviors. In automotive driving intelligent cockpit interaction occupies the visual and tactile sense has a greater impact on driving behavior. Eric pointed out that in the driving process should not over occupy the tactile channel and visual pass band, such as gesture interaction [12], touch interaction, etc., will obviously enhance the safety risk of driving. Wu showed that the automotive intelligent cockpit should give priority to the use of voice interaction [13]. Fatigue driving is also the main cause of traffic accidents, driving fatigue function model [14]. Through the main factor analysis, and finally screened out the five major indicators of the driver's responsiveness, balance stability, visual function, attention ability, speed judgment ability, to establish a multi-factor model for driver fatigue evaluation. The interaction forms of different channels also have different effects on fatigue, and various sources of arousal, such as emotions, drugs, muscle tension, and strong stimulation will affect the resource allocation scheme.

User Experience

Hutchins [15] categorizes automotive UX into three more detailed "categories-vehicle UX", "cockpit UX" and "total vehicle UX". Chen F [16] discusses the application of the five elements of user experience in automotive interaction design and the advantages and disadvantages of automotive interaction modes from the study of user experience in automotive interaction design theory.

In summary cognitive psychology, ergonomics and user experience are the most common evaluation indexes in assessing driver interaction [17]. Among them, cognitive psychology focuses on attention competition (Interference Index), human-computer ergonomics focuses on the operational efficiency of interaction and driving behavior

(Effectiveness index), and user experience is finally expressed as user preference (Preference index). The effects of interaction and automatic driving technology on driving (Help Index) have a comprehensive effect on driving behavior [18].

3 Materials and Methods

3.1 Designer Interviews

For the purpose of this study, we conducted semi-structured interviews with interaction designers in car companies. We chose 5 interaction designers (traditional car companies) and 5 Internet driving business product managers (new car companies) from FAW Group. The interviews summarize the problems encountered in the research 1. to get to know their difficulties in the current design process and their views on the development of automotive interaction in the self-driving scenario. 2. to test the practicality and ease of use of the linear design diagrams output from the designers' research related to the field of automotive interaction. 3. The dimensions and indexes for evaluating the design of automotive interactions in the future.

The outline of the interview is shown in the Fig. 3.

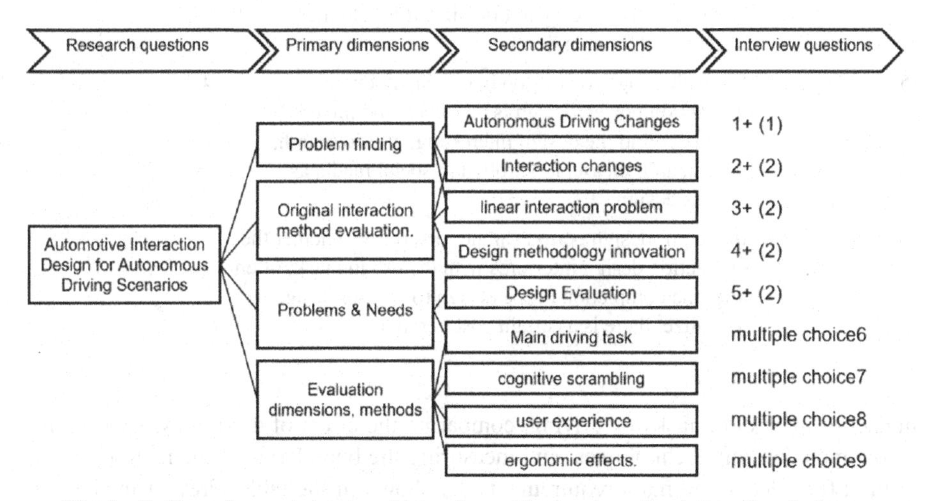

Fig. 3. Outline of a semi-structured interview with an automotive interaction designer.

We summarize the main concerns and feedback of designers through the clustering and correlation method. The following Table describes the main results of the qualitative questions in Table 2.

Interaction designers have suggested that interaction design at this stage encounters a bottleneck, which is unable to balance the relationship between interaction and driving, and there is no suitable tool to deduce the relationship between driving tasks and interaction. We explore how to redefine the human-vehicle interaction relationship outside of the original framework and produce innovative interaction design methods for human-vehicle cooperation in the autonomous driving environment (L0-L5). Focusing on the

Table 2. Summary of views from interviews.

Question Feedback		Number
1	In the auto driving scenario, the form of interaction changes, designers can not only focus on the interior interaction, but also to make efforts in the exterior interaction	6
2	Nowadays, automobile interaction design is undergoing great changes, and it is better not to balance the migration of common design tools in automobile design scenarios. It is possible to jump out of the original design method. Automotive scenarios need new interaction design methods as a guide	6
3	In the future autonomous driving era, the car will be transformed into the third living space. In-vehicle interactions are carried out entirely with people	5
4	With the increase of functions and diversification of interaction, the linear interaction form has too many lines, too chaotic relationships, and is not easy to organize. It is not easy to organize. It is seldom applied in actual interaction design	4
5	Car technology focuses on intelligent cockpit interaction power, design is too much and mixed. Many designers find the design interesting, after research, design, and release, it is trolled in social media as useless design	4
6	In the pre-design period can not determine whether the interaction design interferes with the driving task, often after the completion of the design to do user designers to realize, there is a certain cost iterative	3

impact of interaction tasks on driving, comparing the effect of driving assistance with autonomous driving technologies, and measuring the boundaries of interaction design in the L0-L5 level, the main requirements are shown in the table. Prevent interaction redundancy, which leads to an imbalance between interactivity and safety, as Table 3.

Table 3. Designer Issues Needs Spotlights.

Demand Focus		number
1	New Interaction Design Methods	10
2	The relationship between technical assistance, interaction, and driving	10

3.2 The Concept of Interaction Sets

To address these issues, we introduce the concept of interacting sets. A set is a fundamental concept in mathematics and the main object of study in set theory. The basic theory of set theory was founded in the 19th century, and the simplest statement about a set is the definition in plain set theory (the most primitive set theory), that is, a set is a "definite pile of things", and the "things" in a set are called elements. A set is a collection of concrete or abstract objects of a particular nature, called the elements of the set, and a set of numbers is a collection of numbers. The scope of the set is larger than that of the number set, which is only one kind of set. There are four basic operations on sets, namely intersection, union, difference, complement, and whether an element belongs to a set.

Introducing sets we can categorize and perform basic calculations on driving related elements. Intersection sets are the same as number sets, having elements, subsets, and sets, and can perform operations such as $M \cup N$, $M \cap N$, $N \subseteq M$. By establishing the computation of driving and interaction tasks for autonomous driving technology, we can get more accurate and intuitive information about the feasibility of interaction design and what kind of interactions do not interfere with driving in different autonomous driving level scenarios. We can better sort out the relationship between each task and the scenario, and find the center of this scenario. We can not only show the connection between driving tasks and interaction tasks, but also determine their competition and encroachment with each other.

We define the driving task set as N_{drive}, the driving tasks are mainly composed of driver driving and automated driving. The concatenation of the set of driver tasks (H_{drive}), the set of automated driving tasks (V_{drive}) comprises the set of driving tasks. The driver task (H_{tesk}) consists of the concatenation of the driver interaction task (H_{hmi}) and the driver driving task (H_{drive}), which is directly related to the driver's attention, cognition, physical strength, and audio-visual channel occupancy.

$$H_{drive} \cup V_{drive} = N_{drive}, H_{hmi} \cup H_{drive} = H_{tesk}$$

We represent the aggregated case of the autopilot scenario hierarchy as Table 4.

With such an ensemble representation we can clearly see the relationship between vehicle traveling, human driving tasks, automated driving tasks, and smart cockpit interactions, and the elements and effects of the automated driving tasks are often known in the design of the car company. In order to ensure that the vehicle traveling is not interfered, the interaction tasks can be designed as long as they are within the scope of the set (H_{hmi}). It is represented by the set as:

$$H_{hmi} = H_{tesk} - H_{drive} = N_{drive} - V_{drive}$$

Table 4. Derivation of a Ring Interaction System for Autonomous Driving Scenarios

Level	Driving Mission Collection	Known elements (technical elements)
L0	H_{drive}	
L1	H_{hmi} H_{drive} V_{drive}	V_{drive}={(ABS),(ESP),}
L2	H_{hmi} H_{drive} V_{drive}	V_{drive}={(ACC),(ABD),(APS),(ABS),...}
L3	H_{hmi} H_{drive} V_{drive}	V_{drive}={(NGP),(TJP),LIDAR,...}
L4	H_{hmi} V_{drive} H_{drive}	V_{drive}={Sensing Layer,Decision-making layer,Execution layer,...}
L5	H_{hmi} V_{drive} H_{drive}	

3.3 Questionnaire Survey

The above set is still a representation of nature and relationship, we want to establish a representation of quantitative relationship, in order to seek the quantitative relationship of the set see, we invite designers and drivers to conduct a questionnaire survey. Evaluation dimensions and evaluation indexes are established together. We collected 10 designers, 150 users questionnaires, of which 139 valid questionnaires.

In the literature research, we found that attention contention (Interference Index), human-computer ergonomics is mainly concerned with the efficiency of interaction behavior and driving behavior (Effectiveness index), and the user experience is finally expressed as the user's preference (Preference index). These are the aspects that interaction designers often focus on, but each of them has its own percentage. We would like to summarize the proportion of designers who think they play a role in the interaction and the proportion of each item in Table 5.

Table 5. Average of the share of interaction elements

Evaluation indicators	Scramble for attention X1	Human-computer ergonomics X2	User experience X3
Perceived impact	37%	10%	78%

On this basis, we take a brand of automobile entertainment interaction as an example, and investigate the user's demand index for each interaction task, the interference index of the task and the effect of the role. The demand index is the actual time that users are willing to use this interaction during 30 min of driving. The cognitive index is the percentage of attention that the user evaluates during the use of the interaction compared to the driving of the car. Ergonomics for whether completing this item was helpful or difficult for the driving behaviorist, with a 1 + n% or 1−n% assisting effect.The results are presented in Table 6.

Table 6. Categorization and Evaluation of Entertainment Interactions

Entertainment elements (x|x = n)	Demand Index n1	Cognitive struggle Index n2	Help/Interference Index n3
n_1 = Games	0.30	0.90	−0.36
n_2 = Music	0.83	0.83	0.10
n_3 = Movies	0.66	0.98	−0.25
n_4 = Broadcasting	0.78	0.30	−0.10
n_5 = Social Media	0.77	0.10	−0.15
n_6 = Virtual Reality	0.93	0.56	0.02
n_7 = Singing	0.35	0.20	−0.05

In the table is shown the average of the questionnaire scores, again based on which we weighted the parameters according to cognitive contention and user experience tendencies. Then the effect of ergonomics was assigned and finally the effect of each interaction on the driving task was derived. The overall final result of the entertainment interaction of this car is −0.2292, which means that the overall performance of entertainment on the driving task is an encroachment of 22.92%, the results are presented in Table 7.

$$H_e = \{y|y = h_1, h_2, h_3, h_4, \ldots \in H_{hmi}\},$$
$$h = (x_1 \times n_1 + x_2 n_1) x_3 n_3 \in H_{hmi}$$
$$H_e = h_1 + h_2 + h_3 + h_4 \ldots$$

4 Results and Visualization

4.1 Entertainment System Visualization

The area of the driver task is set to 1. The area of a particular entertainment interaction is defined in terms of a demand index, e.g., the area of the set of game interactions is 0.3. The area of the final interaction set in terms of the demand intrusiveness index is H_{hmi} area. For example, the area of the intersection of game interactions is 0.1908, and the combined performance is the intrusion, i.e., subtraction, reasoning that all sets of recreational interactions are shown in Fig. 4.

Table 7. Weighted Recreational Interaction Calculations

Entertainment elements (x|x = n)	H_{hmi}	
n_1 = Games	−0.1908	−0.2292
n_2 = Music	0.161	
n_3 = Movies	−0.097	
n_4 = Broadcasting	−0.048	
n_5 = Social Media	−0.107	
n_6 = Virtual Reality	0.0716	
n_7 = Singing	−0.019	

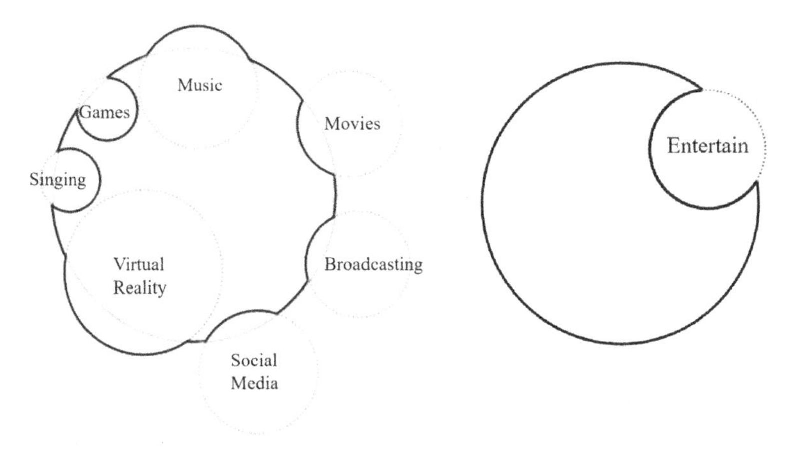

Fig. 4. Automotive Entertainment Interactive Collection Design

The formula for the aggregate area of each term is as follows:

$$S_{h\ task=1}$$
$$S_{games1} = 0.30, S_{music1} = 0.83, S_{movie1} = 0.66, S_{broadcasting1} = 0.78,$$
$$S_{social\ media1} = 0.77, S_{virtual\ reality1} = 0.93, S_{sing1} = 0.35$$
$$S_i = S_{games} - H_{drive}$$
$$S_{games} = -0.19, S_{music} = 0.16, S_{movie} = -0.09, S_{broadcasting} = -0.04,$$
$$S_{cocial\ media} = -0.10, S_{virtual\ reality} = 0.07, S_{sing} = -0.01$$

We aggregate the various entertainment interaction sets together to get the entertainment interaction set about the car and its visual representation in Fig. 4.

$$S_{drive} = 1 - (S_{games} + S_{games} + S_{music}S_{movie}S_{broadcasting} + S_{social\ media} + S_{virtual\ reality} + S_{sing})$$

4.2 Ring Interactive System Design

The percentage of cognitive influence, entertainment influence, and human-computer effect is determined by the designer's evaluation; the interference index, usage preference index, and efficiency index are determined by user testing. Such a tool not only clearly shows the connection between designs, but also demonstrates whether this interaction assists or interferes with the main driving task.

h_{hmi} = (Percentage of cognitive impact × Disturbance Index + Percentage of recreational impact × Usage Preference Index) Percentage of human-computer effect impact × Efficiency Index. This is the canonical formula for evaluating left and right metrics for our interaction set tool in Fig. 5).

$$h = (x_1 \times n_1 + x_2 n_1) x_3 n_3 \in \mathrm{H}_{hmi}$$

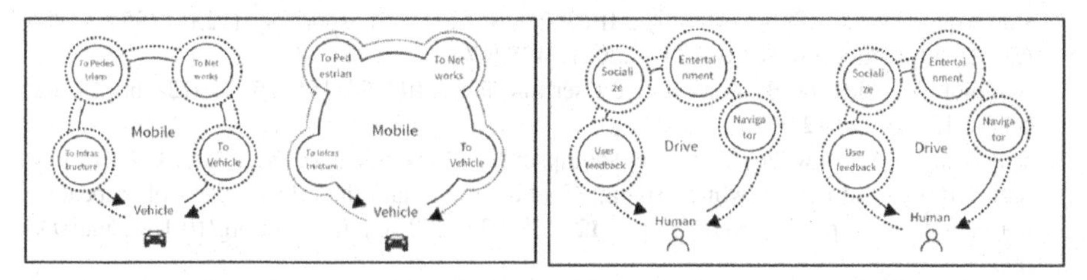

Fig. 5. Human-centric smart cockpit interaction and car-centric driving

In the future L5 level autonomous driving, we can establish the dual ring interaction and of the autonomous driving scenario. They are human-centered intelligent cockpit ring interaction set, and vehicle-centered automated driving ring interaction set. In the dual-loop interaction coupling subset focuses on enhancing user automation trust and driving experience. The interaction design will not be constrained by the driving task. Autonomous driving defines the transformation of the car's interior cabin into a third living space through immersive environments, personalized functions, living scenarios, and information-based services, such as Fig. 6.

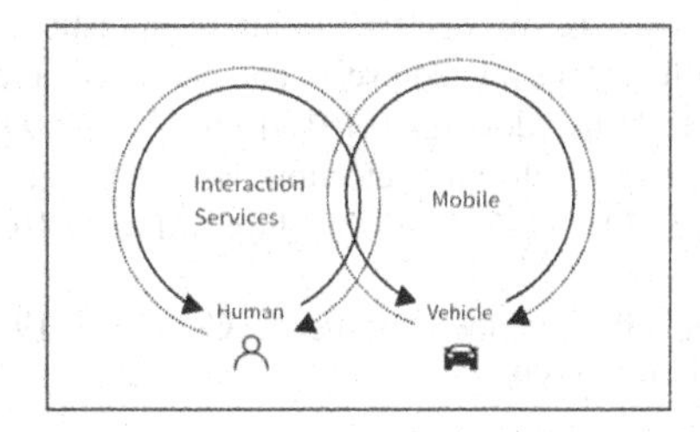

Fig. 6. Autonomous Driving Dual Ring Interaction Set in L5

References

1. Sae, T.: Definitions for terms related to driving automation systems for on-road motor vehicles. SAE Stand. J. (2016)
2. Rosener, C., Sauerbier, J., Zlocki, A., et al.: A comprehensive evaluation approach for highly automated driving. In: 125th international Technical Conference on the Enhanced Safety of Vehicles (ESV). Detroit, Michigan, U.S.A.: NHTSA, pp. 1–13 (2017)
3. Dang, R., Wang, J., Li, S.E., Li, K.: Coordinated adaptive cruise control system with lane-change assistance. IEEE Trans. Intell. Transport. Syst. **16**, 2373–2383 (2015). https://doi.org/10.1109/TITS.2015.2389527
4. Zhao, X., Sun, J., Wang, M.: Measuring sociality in driving interaction. IEEE Trans. Intell. Transport. Syst., 1–14 (2024). https://doi.org/10.1109/TITS.2024.3383867
5. Yang, L., Liu, S.: Analysis on the development status of ICV. In: 2021 International Wireless Communications and Mobile Computing (IWCMC), pp. 2153–2156. IEEE, Harbin City, China (2021). https://doi.org/10.1109/IWCMC51323.2021.9498585
6. Tan, Z., et al.: Human-machine interaction in intelligent and connected vehicles: a review of status quo, issues, and opportunities. IEEE Trans. Intell. Transport. Syst. **23**, 13954–13975 (2022). https://doi.org/10.1109/TITS.2021.3127217
7. Schmidt, A., Pfleging, B.: Automotive User Interfaces. ItIT **54**, 155–156 (2012). https://doi.org/10.1524/itit.2012.9076
8. Sanderson, J.A., Bowden, V., Swire-Thompson, B., Lewandowsky, S., Ecker, U.K.H.: Listening to misinformation while driving: Cognitive load and the effectiveness of (repeated) corrections. J. Appl. Res. Mem. Cogn. **12**, 325–334 (2023). https://doi.org/10.1037/mac0000057
9. Faas, S.M., Kraus, J., Schoenhals, A., Baumann, M.: Calibrating pedestrians' trust in automated vehicles: does an intent display in an external HMI support trust calibration and safe crossing behavior? In: Proceedings of the 2021 CHI Conference on Human Factors in Computing Systems, pp. 1–17. ACM, Yokohama Japan (2021). https://doi.org/10.1145/3411764.3445738
10. Zhao, Y., et al.: EEG assessment of driving cognitive distraction caused by central control information. Presented at the (2023). https://doi.org/10.54941/ahfe1003011
11. Chen, S., Jian, Z., Huang, Y., Chen, Y., Zhou, Z., Zheng, N.: Autonomous driving: cognitive construction and situation understanding. Sci. China Inf. Sci. **62**, 81101 (2019). https://doi.org/10.1007/s11432-018-9850-9
12. (Eric) Li, Y., Hao, H., Gibbons, R.B., Medina, A.: Understanding gap acceptance behavior at unsignalized intersections using naturalistic driving study data. Transp. Res. Record. **2675**, 1345–1358 (2021). https://doi.org/10.1177/03611981211007140
13. Wu, K.-F., Aguero-Valverde, J., Jovanis, P.P.: Using naturalistic driving data to explore the association between traffic safety-related events and crash risk at driver level. Accid. Anal. Prev. **72**, 210–218 (2014). https://doi.org/10.1016/j.aap.2014.07.005
14. Jia, H., Xiao, Z., Ji, P.: Fatigue driving detection based on deep learning and multi-index fusion. IEEE Access. **9**, 147054–147062 (2021). https://doi.org/10.1109/ACCESS.2021.3123388
15. Hutchins, E.: How a cockpit remembers its speeds. Cogn. Sci. **19**, 265–288 (1995). https://doi.org/10.1207/s15516709cog1903_1

16. Chen, F., Terken, J.: Automotive Interaction Design: From Theory to Practice. Springer Nature Singapore, Singapore (2023). https://doi.org/10.1007/978-981-19-3448-3
17. Marafie, Z., et al.: AutoCoach: an intelligent driver behavior feedback agent with personality-based driver models. Electronics **10**, 1361 (2021). https://doi.org/10.3390/electronics10111361

Effects of Automated Vehicles' Transparency on Trust, Situation Awareness, and Mental Workload

Weixing Huang, Milei Chen, Weitao Li, and Tingru Zhang(✉)

Institute of Human Factors and Ergonomics, College of Mechatronics and Control Engineering, Shenzhen University, Shenzhen, China
2110292026@email.szu.edu.cn

Abstract. Understanding the behavioral intentions and decision-making mechanisms of automated vehicles (AVs) is one of the reasons why people trust them. This study aims to explore the main effect of AV transparency on trust, situation awareness and mental workload, while considering whether different driving experiences and scenario types have a moderating effect on the above results. This study conducted a driving simulation experiment. The experiment adopted a three-factor mixed design with AV transparency (No Explanation, What-only, Why-only, What+Why) and driving experience (Not Rich, Rich) as between-subject variables, and scenario type (Expected, Unexpected) as within-subject variables. To balance the impact of the scenario presentation order in the experiment, half of the participants were exposed to the scenarios in order from expected to unexpected, and the other half were exposed to the scenarios in order from unexpected to expected. Froty-eight participants (24 females and 24 males) participated in the experiment. The research found that increased transparency information help people better perceive and understand AVs while having less workload. In addition, the research found that drivers with rich driving experience are less dependent on the information provided by the AVs. The study also found that drivers have higher trust and lower anxiety in AVs in expected scenarios.

Keywords: Automated Vehicle · Trust · Transparency · Situation Awareness · Mental Workload

1 Introduction

With the rapid development of theory systems, hardware equipment, and artificial intelligence (AI) algorithms, the level of automated vehicles (AVs) intelligence is getting higher and higher. In order to better classify AVs, the Society of Automotive Engineers (SAE) has proposed a widely accepted classification standard (Committee, 2021), ranging from Level-0 (No Driving Automation) to Level-5 (Full Driving Automation). Restricted by regulations, ethics, society, etc., most of the existing AVs are at Level-2 or Level-3. Under these two levels, although a certain degree of automation has been achieved, the monitoring and intervention of human drivers are still required to ensure

H. Krömker (Ed.): HCII 2024, LNCS 14732, pp. 116–132, 2024.
https://doi.org/10.1007/978-3-031-60477-5_9

safety and reliability. To this end, relevant researchers are actively exploring the introduction of advanced AI algorithms to enable real-time analysis and decision-making on complex driving scenarios, thereby making AVs safer and more reliable in various environments and traffic conditions. However, the black-box nature and non-transparency of AI algorithms cause people to have inappropriate trust in AVs (Faber & van Lierop, 2020; Holländer et al., 2019), which greatly affects people's use of AVs, and is also not good for the development of human-machine co-driving.

Trust is a determinant of people's use of automated systems (Schaefer et al., 2016). In human-machine systems, trust can be defined as "the attitude that an agent will help achieve an individual's goals in a situation characterized by uncertainty and vulnerability" (Lee & See, 2004). It can help people overcome the risks and uncertainties in operating automated systems to make decisions and take corresponding actions when they may need to bear negative consequences and risks (Yan et al., 2011). People's over-trust or distrust in automated systems will affect their interaction of automated systems and make them unable to complete tasks normally, even threatening people's safety in some cases. Therefore, in the process of people using AVs, how to make people have an appropriate level of trust in AVs has become a current research hot topic.

Previous research has shown that increasing automation transparency enables people to use automation more accurately and efficiently (Bhaskara et al., 2021; van de Merwe et al., 2022). Transparency means the degree to which the internal workings or mechanisms of the automated systems are known to humans (Seong & Bisantz, 2008). In recent years, theoretical models of transparency have emerged one after another. In human-machine co-driving research, the most used transparency model is the What-Why model (Koo et al., 2015). This model provides the driver with the current behavioral intention or decision-making mechanism of the AV systems by providing transparency information on the "what" type or the "why" type. In Koo et al. (2015) study, providing "why" information was the drivers' first choice and also led to good driving performance, while providing "how+why" information resulted in the safest driving performance. In terms of trust, research shows that increasing transparency information about AVs will increase driver trust in AVs (Bhaskara et al., 2020; Tatasciore & Loft, 2024). However, current research on the transparency of AV systems is still limited, and the specific information content that needs to be conveyed to user needs to be further clarified. In addition, how to effectively convey this information to people is also a problem that needs to be solved. Too much information may confuse users, while not enough information may not build enough trust.

This study intends to design a human-machine co-driving simulation experiment with different transparency based on the What-Why model, with the purpose of:

1. To explore the main effect of AV transparency on trust, mental workload, and situation awareness.
2. To explore the dynamic development rules of AV transparency on trust, anxiety, and the usefulness of AV information.
3. Consider whether different driving experiences and scenario types have a moderating effect on the above results.

2 Method

2.1 Participants

48 subjects were recruited in the simulated driving experiment, with 24 being females and 24 males. All the subjects were students from Shenzhen University with valid driver's licenses. They had normal vision or corrected vision and were not colorblind or color weakness. They were in good physical condition and had no recent symptoms such as colds.

2.2 Driving Scenario and NDRT

Driving scenario. A fully enclosed, two-way 50-km long road without traffic lights was developed, with each lane width set to 3.75 m. The simulated vehicle was an SAE Level-3 AV that can complete almost all driving operations independently, but the driver still needs to maintain concentration to deal with possible situations that the AV cannot handle. In addition, when AV issued a takeover request (TOR), the vehicle forcibly switched vehicle control back to the human driver. When conditions were met to resume autonomous driving, the vehicle prompted the subject to switch to automated driving mode. There were 12 events distributed in the driving scenario as shown in Table 1, including 5 expected events, 5 unexpected events, and 2 takeover events.

Non-driving Related Task (NDRT). The NDRT used in the experiment was the visual

Table 1. The events, types, and descriptions of the experiment.[1]

Scenario	Type	Description
Slow down	Expected	The AV on the road, ahead of the tunnel entrance, then AV deceleration through
Speed up	Expected	The AV followed the car on the road. After a certain distance, the car in front changed lanes. There were no other vehicles in the lane where the AV was, so the AV accelerated
Overtaking on the left	Expected	The AV was on the road and the vehicle in front of the AV was moving slowly, so the AV overtaken on the left side
Overtaking on the right	Expected	The AV on the road, the vehicle in front of the lane moved slowly, left congestion, AV right overtaken

(continued)

[1] Expected events referred to events in which the drivers can understand the information provided by AV based on the current perception of the surrounding environment; unexpected events referred to events in which the drivers cannot understand the information provided by AV based on the current perception of the surrounding environment; takeover events referred to events in which control of the vehicle had to be handed over to the driver due to the malfunction in AV.

Table 1. (*continued*)

Scenario	Type	Description
Change lane left	Expected	The AV drove on the ramp, decelerated until it stopped, and waited for an opportunity to change lanes to the left at the entrance ramp
Change lane right	Unexpected	The AV was driving on the road, and the IoT system detected that there was a traffic accident ahead, so the AV changed lanes to the right in advance
Stop when turning right	Unexpected	When AV was driving on the road when it turned right at the T-junction without a traffic signal because there were other vehicles parked on both sides of the T-junction, AV's visual field was blocked, so it stopped and observed. After 5 s, AV continued to complete the right turn
Change route	Unexpected	The AV was driving on the road, and the IoT system detected that the road ahead was temporarily closed to traffic, so the AV changed its route
Slow down and stop	Unexpected	The AV was on the road when it suddenly slowed down and stopped. After 5 s a fire engine honked and passed quickly, and AV continued to drive
Slow down and drive slowly	Unexpected	The AV drove on the road, because the right side of the road vision was blocked by vehicles, in order to prevent a sudden pedestrian across the street, AV slowed down
Blurred road markings	Takeover	The AV drove on the road detected a part of the road ahead of the line marking fuzzy, so the AV let the driver take over the vehicle
Sensor failure	Takeover	The AV on the road, fogs after a period of time, led to the failure of AV sensors, so the AV let the driver take over the vehicle

dual 1-back memory task. Stimuli were randomly generated in 9 (3*3) squares, and participants were asked to compare the current stimulus with the previous stimulus adjacent to it. If the location of the current stimulus was consistent with the location of the previous stimulus, pressed the location consistent button; if the color of the current stimulus was consistent with the color of the previous stimulus, pressed the color consistent button. The NDRT was implemented using a Huawei MatePad Pro tablet (10.8 inches) placed on the right side of the steering wheel, similar to the position of the center control panel of a real vehicle, which allowed the participants to divert their visual attention from the road while performing the NDRT.

2.3 Apparatus

Driving Simulator. The driving simulator consisted of three 27-in. LED displays (with a resolution of 1920 * 1080), a desktop host computer installed with UC-win/Road Ver14.0, a Logitech G29 dual-motor force feedback steering wheel and pedals, and an adjustable driver's seat. While provided a medium-fidelity driving experience, the driving simulator can also collect driving data such as speed and acceleration at a frequency of 60 Hz.

Questionnaire. To collect subjects' subjective evaluations of AVs and the information provided by AVs, we designed a self-administered questionnaire, which was modified based on the existing questionnaires to evaluate the experimental subjects' mental workload (Hart & Staveland, 1988), trust in automation (Körber, 2019), satisfaction (Hoffman et al., 2018), perceived transparency (Shin, 2021), and perceived usefulness (Zhang et al., 2019). In this study, subjects filled out this questionnaire after they completed each type of scenario, twice in total. All items were measured using a 7-point Likert scale ranging from "strongly disagree (=1)" to "strongly agree (=7)".

2.4 Experiment Design

The experiment adopted a 4 (Transparency: No Explanation, What-only, Why-only, What + Why) × 2 (Driving experience: Not Rich, Rich) × 2 (Scenario type: Expected, Unexpected) three-factor mixed design. The between-subject factors were transparency and driving experience, and the within-subject was scenario type. Among them, transparency was achieved by providing behavioral intention information (What) or decision-making mechanism information (Why) of AV in the form of visual and auditory. In terms of driving experience, we defined those with a total driving mileage of more than 2,000 km as having rich driving experience, and the remaining of them as not having rich driving experience. In terms of scenario type, we regarded the combination of 5 expected events and blurred road markings event as the expected scenario, and the combination of 5 unexpected events and sensor failure events as the unexpected scenario. At the same time, to balance the influence of the order of scenario presentation in the experiment, half subjects were experiencing the expected scenario first and then the unexpected scenario, and the other half were experiencing the unexpected first and then the expected scenario.

2.5 Procedure

The process of each event was shown in Fig. 1. In the oral report section, subjects were asked to report their trust in AV, their anxiety in AV, and the usefulness of the information provided by AV on a scale of 0–100. Then they were asked two questions about situation awareness (SA), SA1 asked what action AV performed, and SA2 asked why AV performed the action, each question provided 3 options to choose from. Table 2 took the left overtaking event as an example and presented the questions that subjects needed to oral report after experiencing the event.

Table 2. The questions that subject need to oral report after the left overtaking event.[2]

Item	Question
Trust in AV	Do you trust in AV? (Please rate within 0–100.)
Anxiety in AV	Does trusting AV make you anxiety? (Please rate within 0–100.)
The usefulness of the information provided by AV	How useful is the information provided by AV to you? (Please rate within 0–100.)
SA1	What action AV performed in the event just now? 1-Turn left; 2-Overtaking on the left; 3-Request to takeover
SA2	Why AV performed the action? 1-Traffic accident; 2-Enter the tunnel; 3-The vehicle ahead is slow

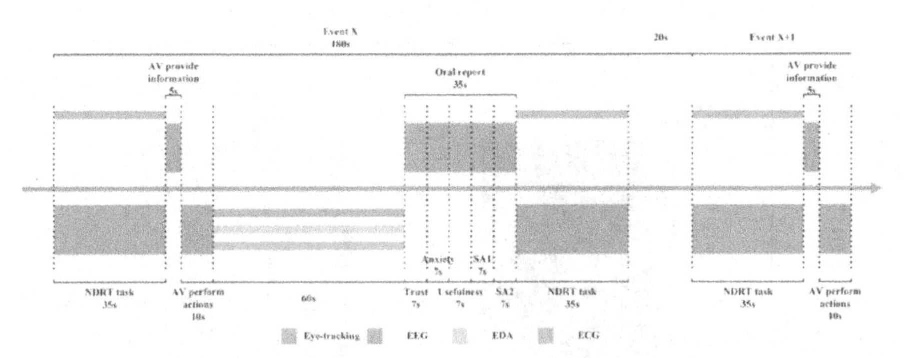

Fig. 1. The process of each event in the experiment.

3 Results

For the three questions of trust in AV, anxiety in AV, and the usefulness of the information provided by AV, we collected the subjects' ratings and then analyzed them using the linear mixed model (LMM). For SA1 and SA2, we calculated the accuracy of the subjects in the expected and unexpected scenarios, respectively. Then we used analysis of variance (ANOVA) to conduct the study. For the questionnaire, we used ANOVA for research.

3.1 Dynamic Measurement

Trust. The LMM results of dynamic trust in the AV were shown in Table 3. The results showed that the interaction effects of two factors were no significant, and the main effect

[2] In SA1 and SA2, the correct answer was underlined.

of scenario type ($F(1,94) = 6.83$, $p = 0.009$) was significant, but the main effects of transparency and driving experience were no significant difference.

Table 3. The LMM results of dynamic trust in the AV.[3]

Variable	Df		F	p
Transparency	3	44	0.27	0.844
Driving Experience	1	46	1.99	0.166
Scenario Types	1	94	6.83	0.009**
Transparency × Driving Experience	3	40	0.79	0.506
Transparency × Scenario Types	3	88	0.76	0.515

The post hoc comparisons of scenario type (Fig. 2) fount that drivers' trust in AVs in expected scenarios were significantly higher than that in unexpected scenario.

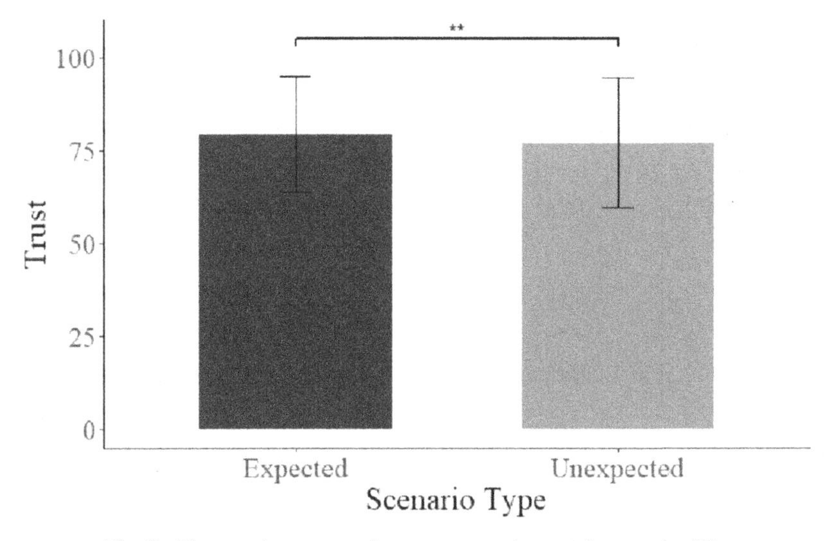

Fig. 2. The post hoc comparisons on scenario type for trust in AVs.

Anxiety. The LMM results of dynamic anxiety in the AV are shown in Table 4. A significant main effect of scenario type ($F(1,94) = 18.12$, $p < 0.001$) on the drivers' anxiety in the AV, and the other factors were no significant.

The post hoc comparisons of scenario type (Fig. 3) found that drivers' anxiety in AVs in expected scenarios were significantly lower than that in unexpected scenarios ($p < 0.001$).

[3] ***p < .001, **p < .01, *p < .05.

Table 4. The LMM results of dynamic anxiety in the AV.

Variable	Df		F	p
Transparency	3	44	0.29	0.831
Driving Experience	1	46	0.33	0.569
Scenario Types	1	94	18.12	<0.001***
Transparency × Driving Experience	3	40	0.43	0.733
Transparency × Scenario Types	3	88	0.41	0.746

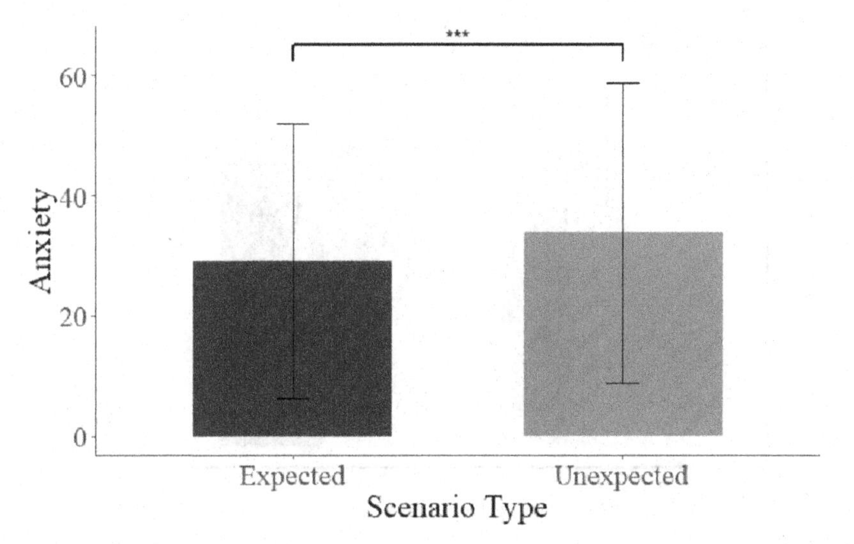

Fig. 3. The post hoc comparisons on scenario type for anxiety in AVs.

Usefulness. The LMM results of the dynamic usefulness of the information provided by the AV were shown in Table 5. The results showed that a significant interaction effect between transparency and driving experience ($F(2, 30) = 3.79$, $p = 0.034$), while there was no significant difference in the other factors.

A simple main effect analysis was conducted on driving experience (Fig. 4). The results showed that drivers with not rich driving experience believed that the usefulness of What-only information provided was significantly greater than that of drivers with rich driving experience ($p = 0.019$). While the simple main effect analysis of transparency showed no significant difference.

SA1 Accuracy. The ANOVA results of the SA1 accuracy are shown in Table 6. The results show that the interaction effect of two factors and three factors are no significant, while the main effects of transparency ($F(3,44) = 7.81$, $p < 0.001$) and scenario type ($F(1,94) = 17.07$, $p < 0.001$) are significant.

Table 5. The LMM results of dynamic usefulness of the information provided by the AV.

Variable	Df		F	p
Transparency	2	33	1.10	0.344
Driving Experience	1	34	0.35	0.558
Scenario Types	1	70	0.01	0.936
Transparency × Driving Experience	2	30	3.79	0.034*
Transparency × Scenario Types	2	66	0.91	0.405

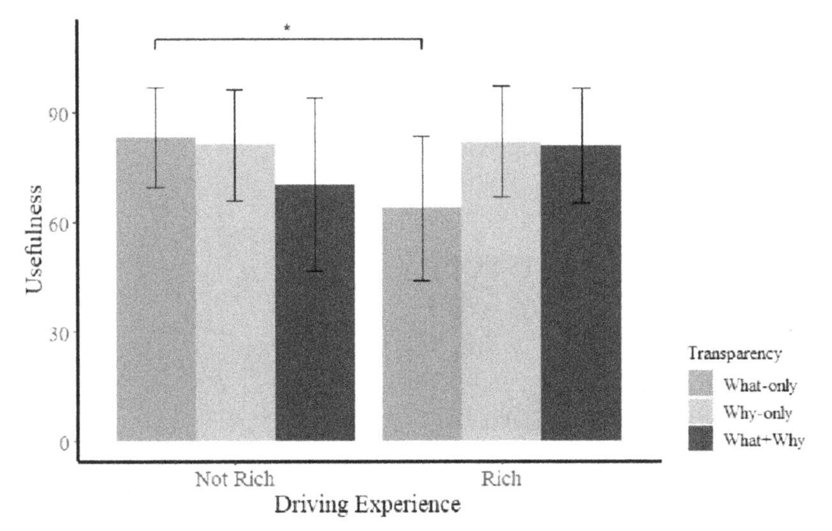

Fig. 4. The interaction effect and post hoc comparisons of transparency and driving experience on usefulness of the information provided by the AV.

Table 6. The ANOVA results of the SA1 accuracy.

Variable	Df		F	p
Transparency	3	44	7.81	<0.001***
Driving Experience	1	46	0.39	0.535
Scenario Types	1	94	17.07	<0.001***
Transparency × Driving Experience	3	40	0.97	0.416
Transparency × Scenario Types	3	88	0.29	0.832

The post hoc comparisons on transparency (Fig. 5) showed that the SA1 accuracy of AVs that provide What information is significantly greatly than that of AVs that do not provide What information.

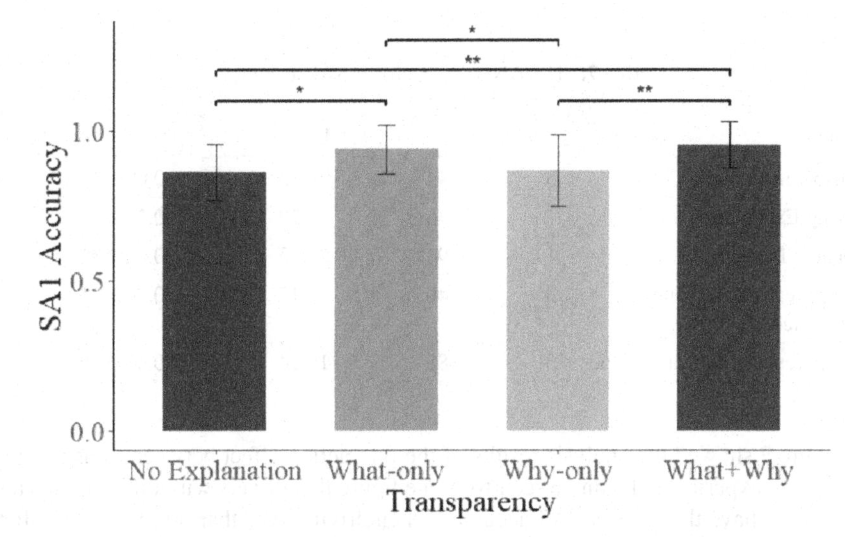

Fig. 5. The post hoc comparisons on transparency for SA1 accuracy.

Figure 6 showed the post hoc comparisons results of scenario type. It can be seen from the figure that the SA1 accuracy of the expected scenario is significantly greater than that of the unexpected scenario.

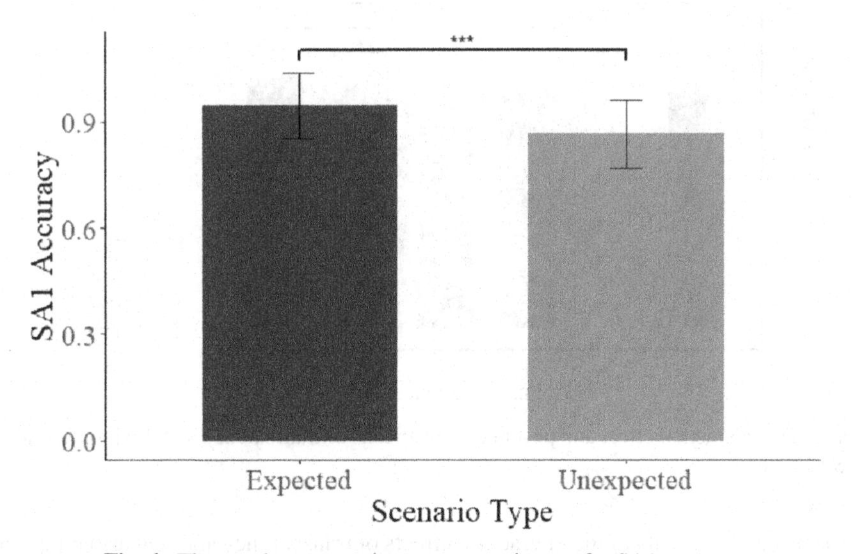

Fig. 6. The post hoc comparisons on scenario type for SA1 accuracy.

SA2 Accuracy. The ANOVA results of SA2 accuracy were shown in Table 7. The results showed that a significant interaction between transparency and driving experience (F (3,40) = 8.17, p < 0.001), a significant interaction effect between transparency and scenario type (F (3,88) = 10.94, p < 0.001).

Table 7. The ANOVA results of SA2 accuracy

Variable	Df		F	*p*
Transparency	3	44	26.13	<0.001***
Driving Experience	1	46	1.29	0.263
Scenario Types	1	94	29.65	<0.001***
Transparency × Driving Experience	3	40	8.17	<0.001***
Transparency × Scenario Types	3	88	10.94	<0.001***

Figure 7 showed the analysis results of the interaction effect between transparency and driving experience. It can be seen from the figure that drivers with different driving experiences have the highest SA2 accuracy when driving AVs that provide Why information. Specifically, drivers had significantly higher SA2 accuracy when driving AVs that provided Why information than AVs that provided What-only information.

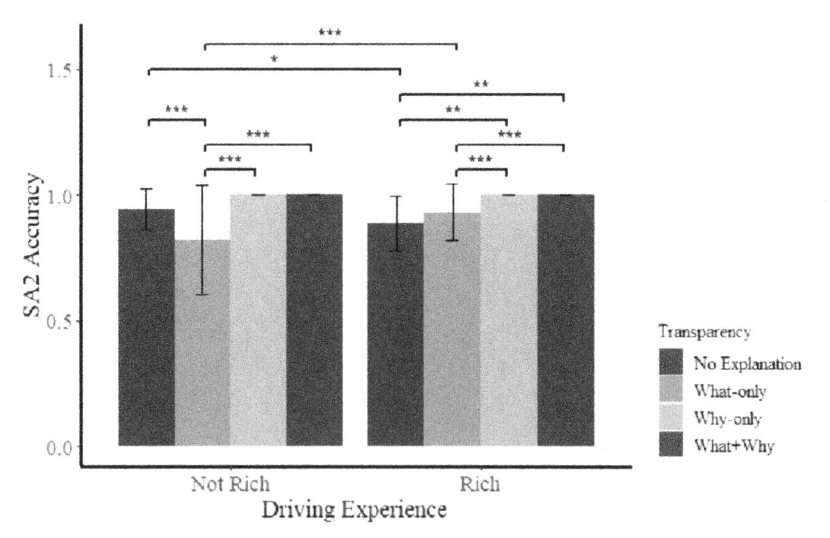

Fig. 7. The interaction effect and post hoc comparisons of transparency and driving experience on SA2 accuracy.

The analysis results of the interaction effects of transparency and scenario type were shown in Fig. 8. As can be seen from the figure, the SA2 accuracy of AVs that provide

Why information in unexpected scenarios was significantly greater than that of AVs that not provide Why information.

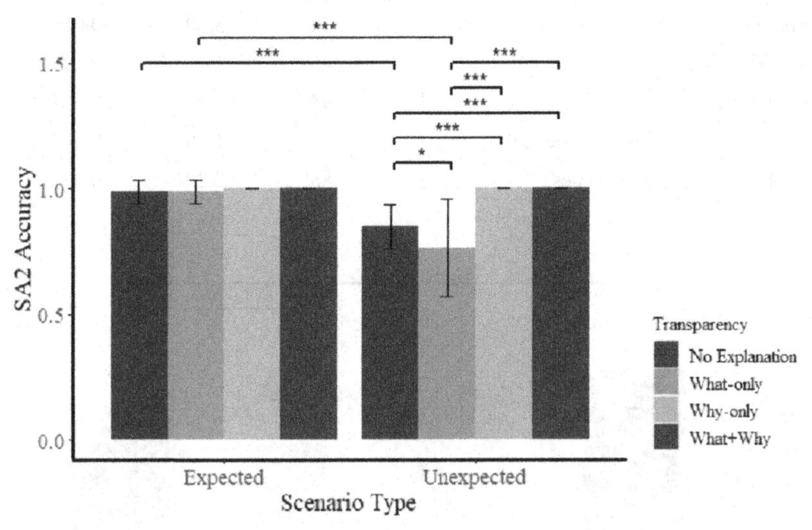

Fig. 8. The interaction effect and post hoc comparisons of transparency and scenario type on SA2 accuracy

3.2 Questionnaire Analysis

Trust in Automation. For trust in automation, the results are shown in Table 8, and there was no significant difference.

Table 8. The ANOVA results of trust in automation.

Variable	Df		F	*p*
Transparency	3	44	0.19	0.902
Driving Experience	1	46	0.36	0.552
Scenario Types	1	94	0.78	0.383
Transparency × Driving Experience	3	40	0.76	0.523
Transparency × Scenario Types	3	88	0.25	0.863

Mental Workload. For mental workload, the ANOVA results (Table 9) showed that the main effect of transparency was significant ($F(3,44) = 3.60$, $p = 0.017$), and post hoc comparisons (Fig. 9) revealed that the mental workload of AVs that provide What + Why information is significantly less than that of no explanation ($p = 0.046$) and provide Why-only information ($p = 0.023$).

Table 9. The ANOVA results of mental workload.

Variable	Df		F	*p*
Transparency	3	44	3.60	0.017*
Driving Experience	1	46	1.69	0.198
Scenario Types	1	94	0.28	0.601
Transparency × Driving Experience	3	40	1.86	0.143
Transparency × Scenario Types	3	88	0.29	0.834

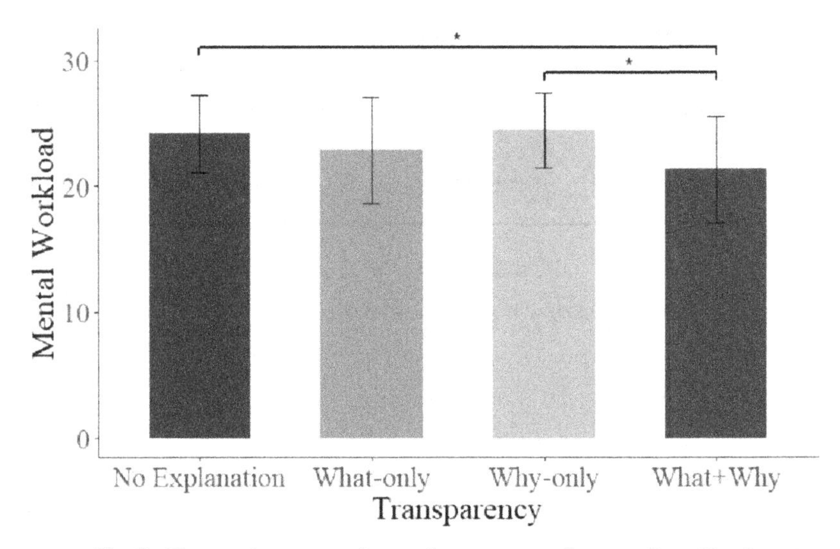

Fig. 9. The post hoc comparisons of transparency for mental workload.

Situation Awareness. The ANOVA results of situation awareness are shown in Table 10, and there was no significant difference.

Table 10. The ANOVA result of situation awareness.

Variable	Df		F	*p*
Transparency	3	44	0.65	0.590
Driving Experience	1	46	0.39	0.539
Scenario Types	1	94	0.20	0.660
Transparency × Driving Experience	3	40	0.52	0.674
Transparency × Scenario Types	3	88	0.75	0.528

Satisfaction. The results of satisfaction are shown in Table 11. Since this part is to measure satisfaction with the information provided by AVs, participants with no explanation AVs will not be involved in this part of the questionnaire. As shown in the table, there was no significant difference.

Table 11. The ANOVA results of satisfaction.

Variable	Df		F	p
Transparency	2	33	0.29	0.752
Driving Experience	1	34	0.67	0.420
Scenario Types	1	70	1.64	0.210
Transparency × Driving Experience	2	30	0.61	0.551
Transparency × Scenario Types	2	66	0.13	0.882

Perceived Transparency. Perceived transparency also, like satisfaction, excluded participants with no explanation AVs. The results of perceived transparency show that there was no significant difference (Table 12).

Table 12. The ANOVA results of perceived transparency.

Variable	Df		F	p
Transparency	2	33	0.11	0.898
Driving Experience	1	34	0.17	0.685
Scenario Types	1	70	0.53	0.472
Transparency × Driving Experience	2	30	0.16	0.856
Transparency × Scenario Types	2	66	0.86	0.435

Perceived Usefulness. The same as satisfaction and perceived transparency, perceived usefulness excluded participants with no explanation AVs. The ANOVA results of perceived usefulness are shown in Table 13, and there was no significant difference.

Table 13. The ANOVA results of perceived usefulness.

Variable	Df		F	p
Transparency	2	33	0.01	0.988
Driving Experience	1	34	0.88	0.356
Scenario Types	1	70	0.00	0.951
Transparency × Driving Experience	2	30	1.03	0.369
Transparency × Scenario Types	2	66	0.75	0.481

4 Discussion

4.1 Dynamic Measurement Analysis Results

The analysis results of trust in AVs show that driver's trust in AV in the expected scenarios is significantly higher than that in the unexpected scenarios. It can be considered that drivers will trust in AVs more in expected scenarios than in unexpected scenarios. The analysis results of anxiety in AVs show that driver's anxiety in AVs in the expected scenarios is significantly lower than that in the unexpected scenarios. This is somewhat different from research results of Stephenson et al. (2020). In their research, there was no difference in drivers' trust between the two after experiencing expected and unexpected scenarios, but the anxiety represented by electrodermal signals increased.

The analysis results of the usefulness of the information provided by AVs show that for AVs that provide What-only information, drivers with not rich driving experience think that the usefulness of the information they provide is significantly higher than that of drivers with rich driving experience. It may be caused by drivers with not rich driving experience relying more on AVs.

4.2 Situation Awareness Accuracy

The SA1 accuracy research results show that the SA1 accuracy when the driver drives the AVs that provide What information is significantly higher than that that does not provide What information. It can be considered that providing the behavioral intention information of the AVs helps the drivers better understand the next steps of the action to be performed by the AVs. In addition, the research results also found that the SA1 accuracy when the driver experienced expected scenarios was significantly higher than that when experiencing unexpected scenarios. This shows that the driver is more aware of the AVs' behavioral intentions in expected scenarios.

The research results of SA2 accuracy show that no matter what kind of driving experience drivers have, driving AVs that provide Why information will have higher SA2 accuracy. In other words, AVs that provide information about decision-making mechanisms can help drivers better understand why the AVs take the appropriate actions. In addition, the SA2 accuracy of AVs that provide Why information when drivers drive in unexpected scenarios is significantly higher than that provide no explanation and

What-only information. This means that providing information about the AVs' decision-making mechanism in unexpected scenarios can help drivers better understand why the AVs take the actions.

4.3 Mental Workload

The mental workload analysis results show that the drivers' mental workload when driving the AVs that provide What+Why information is significantly lower than that driving the AVs that provide no explanation or What-only information. This shows that providing information about the behavioral intentions and decision-making mechanisms of AVs can enable drivers to reduce their workload while driving.

5 Conclusion

The main conclusions of the present study are as follows:

1. Increased transparency information help people better perceive and understand AVs while having less workload.
2. Experienced drivers are less dependent on the information provided by the AVs.
3. Drivers have higher trust and lower anxiety in AVs in expected scenarios.

Acknowledgments. This work was supported by the Guangdong Basic and Applied Basic Research Foundation (grant number 2021A1515011610) and the Foundation of Shenzhen Science and Technology Innovation Committee (grant number JCYJ20210324100014040).

References

Bhaskara, A., et al.: Effect of automation transparency in the management of multiple unmanned vehicles. Appl. Ergon. **90**, 103243 (2021). https://doi.org/10.1016/j.apergo.2020.103243

Bhaskara, A., Skinner, M., Loft, S.: Agent transparency: a review of current theory and evidence. IEEE Trans. Hum.-Mach. Syst. **50**(3), 215–224 (2020). https://doi.org/10.1109/THMS.2020.2965529

Committee, O.-R.A.D.: Taxonomy and definitions for terms related to driving automation systems for on-road motor vehicles. In: SAE International (2021)

Faber, K., van Lierop, D.: How will older adults use automated vehicles? Assessing the role of AVs in overcoming perceived mobility barriers. Transp. Res. Part A Pol. Pract. **133**, 353–363 (2020). https://doi.org/10.1016/j.tra.2020.01.022

Hart, S.G., Staveland, L.E.: Development of NASA-TLX (Task Load Index): results of empirical and theoretical research. In: Hancock, P.A., Meshkati, N. (eds.) Advances in Psychology, vol. 52, pp. 139–183. North-Holland (1988). https://doi.org/10.1016/S0166-4115(08)62386-9

Hoffman, R.R., Mueller, S.T., Klein, G., Litman, J.: Metrics for explainable AI: challenges and prospects. arXiv preprint arXiv:1812.04608 (2018)

Holländer, K., Wintersberger, P., Butz, A.: Overtrust in external cues of automated vehicles: an experimental investigation. In: Proceedings of the 11th International Conference on Automotive User Interfaces and Interactive Vehicular Applications, Utrecht, Netherlands (2019). https://doi.org/10.1145/3342197.3344528

Koo, J., Kwac, J., Ju, W., Steinert, M., Leifer, L., Nass, C.: Why did my car just do that? Explaining semi-autonomous driving actions to improve driver understanding, trust, and performance. Int. J. Interact. Des. Manuf. (IJIDeM) **9**(4), 269–275 (2015). https://doi.org/10.1007/s12008-014-0227-2

Körber, M.: Theoretical considerations and development of a questionnaire to measure trust in automation. In: Bagnara, S., Tartaglia, R., Albolino, S., Alexander, T., Fujita, Y. (eds.) IEA 2018. AISC, vol. 823, pp. 13–30. Springer, Cham (2019). https://doi.org/10.1007/978-3-319-96074-6_2

Lee, J.D., See, K.A.: Trust in automation: designing for appropriate reliance. Hum. Factors **46**(1), 50–80 (2004). https://doi.org/10.1518/hfes.46.1.50_30392

Schaefer, K.E., Chen, J.Y.C., Szalma, J.L., Hancock, P.A.: A meta-analysis of factors influencing the development of trust in automation: implications for understanding autonomy in future systems. Hum. Factors **58**(3), 377–400 (2016). https://doi.org/10.1177/0018720816634228

Seong, Y., Bisantz, A.M.: The impact of cognitive feedback on judgment performance and trust with decision aids. Int. J. Ind. Ergon. **38**(7), 608–625 (2008). https://doi.org/10.1016/j.ergon.2008.01.007

Shin, D.: The effects of explainability and causability on perception, trust, and acceptance: implications for explainable AI. Int. J. Hum.-Comput. Stud. **146** (2021). https://doi.org/10.1016/j.ijhcs.2020.102551

Stephenson, A.C., et al.: Effects of an unexpected and expected event on older adults' autonomic arousal and eye fixations during autonomous driving [original research]. Front. Psychol. **11** (2020). https://doi.org/10.3389/fpsyg.2020.571961

Tatasciore, M., Loft, S.: Can increased automation transparency mitigate the effects of time pressure on automation use? Appl. Ergon. **114**, 104142 (2024). https://doi.org/10.1016/j.apergo.2023.104142

van de Merwe, K., Mallam, S., Nazir, S.: Agent transparency, situation awareness, mental workload, and operator performance: a systematic literature review. Hum. Fact. 00187208221077804 (2022). https://doi.org/10.1177/00187208221077804

Yan, Z., Kantola, R., Zhang, P.: a research model for human-computer trust interaction. In: 2011IEEE 10th International Conference on Trust, Security and Privacy in Computing and Communications (2011)

Zhang, T., Tao, D., Qu, X., Zhang, X., Lin, R., Zhang, W.: The roles of initial trust and perceived risk in public's acceptance of automated vehicles. Transp. Res. Part C: Emer. Technol. **98**, 207–220 (2019)

Exploring Emotional Responses to Anthropomorphic Images in Autonomous Vehicle Displays: An Eye-Tracking Study

Cian-Yun Jun and Jo-Yu Kuo(✉)

Department of Industrial Design, National Taipei University of Technology, Taipei, Taiwan
t112588028@ntut.org.tw, jyk@ntut.edu.tw

Abstract. Anthropomorphic images impart human-like features to non-human entities or abstract concepts can enhance user engagement and emotional connections, thereby enriching digital experiences. This subject has garnered significant interest in the realm of human-computer interaction. However, incorporating anthropomorphic images into digital transportation displays, such as navigation apps and in-vehicle displays, is relatively unexplored. With the rise of human-machine interactions in transportation, it is critical to understand how these visual effects influence user experience in urban mobility.

Therefore, this study, involving 33 participants, investigated the impact of anthropomorphic images and media presentation styles (static versus dynamic) on user experience in autonomous taxi services. We employed eye-tracking technology to measure participants' visual engagement and collected subjective experience data using the User Experience Questionnaire (UEQ-S) and the Self-Assessment Manikin (SAM) scale. The results revealed that dynamic presentation styles elicit emotional resonance and visual attention more effectively than static ones. While anthropomorphic images did enhance visual attention to some extent, there was not a strong correlation between this increased attractiveness and the experience of pleasure.

Overall, this research provides empirical evidence of the potential of anthropomorphic images in shaping emotional experiences in digital displays for transportation services. These insights can inform the selection of visual elements for vehicle display design, acknowledging their impact on user perception and decision-making. Future studies can explore a broader range of anthropomorphic features on emotions, to enhance our predictive understanding of emotional experiences in autonomous vehicles.

Keywords: Eye-Tracking · Emotional Perception · Anthropomorphism · Human-Computer Interaction · User Experience

1 Introduction

With the rapid advancement of autonomous driving technology, traditional taxi services and car-sharing platforms are undergoing significant changes, inevitably influencing future business models and service delivery in transportation [1]. The application of this

H. Krömker (Ed.): HCII 2024, LNCS 14732, pp. 133–144, 2024.
https://doi.org/10.1007/978-3-031-60477-5_10

technology not only heightens the necessity of transportation systems but also has the potential to reshape the way people travel in the future. Consequently, human-machine interaction becomes an indispensable aspect of the user experience in autonomous taxis. This paper explores the evolving landscape of transportation services in the context of autonomous driving technology and discusses the pivotal role of human-machine interaction in shaping the experiential aspect of autonomous taxi services.

1.1 Autonomous Taxi Services

Presently, shared services utilizing autonomous driving vehicles are limited to specific regions. NuTonomy made headlines in 2016 by introducing the world's first public trial of a Robotaxi service in Singapore [2]. Since 2020, Apollo Go's autonomous taxi service has garnered over one million trial rides in cities such as Shenzhen, Shanghai, and Beijing [3]. Additionally, in 2022, Waymo secured approval from the California Public Utilities Commission (CPUC) to participate in testing programs, enabling Waymo to offer truly driverless transportation services to the public in California [4].

Contemporary studies confirm the substantial advantages brought about by autonomous driving technology. Simulation experiments reveal that replacing conventional taxis with autonomous vehicles in New York City could reduce average passenger wait times by 29.82% and improve trip success rates by 7.65% [5]. Additionally, autonomous taxis contribute to a 17% reduction in fleet sizes, resulting in a 17% increase in travel time [6]. Moreover, they can mitigate accidents caused by human errors [7]. Compared to human-driven taxis, autonomous taxis exhibit cost-effectiveness and potentially enhanced safety [8].

The studies above underscore the promising role of autonomous taxis as a transformative trend in urban transportation. However, research on human-machine interaction in autonomous taxi services remains relatively limited, especially concerning passengers' emotional experiences. When passengers access pertinent information through in-car panels, the presentation of images and text is critical for the overall quality of the user experience.

1.2 Anthropomorphic Design

Products with anthropomorphic features have a multifaceted impact on consumers. This impact extends from emotional responses to social connections with the product, enhancing the overall sensory experience [9], and triggering positive emotions to increase customer attractiveness [10]. An experiment on the influence of anthropomorphic application images on user attractiveness found that such images foster positive attitudes and elicit favorable emotions, making them more appealing to users [11]. Additionally, studies on service quality suggest that considering intimacy factors through anthropomorphism positively affects the perception of service quality [12].

Nevertheless, anthropomorphism also comes with potential drawbacks. As its degree increases, user preference might decline, with consumers more likely to elicit positive responses when the level of anthropomorphism is low to moderate [13]. While anthropomorphic design can offer users a friendly product experience, designers must be mindful of potential negative emotional responses at higher degrees of anthropomorphism to avoid user discomfort and aversion.

The style of media presentation also plays a crucial role in the effectiveness of anthropomorphism. A visual perception experiment comparing the impact of dynamic versus static presentations on a product website found that 75% of customers preferred dynamic presentation of visual content [14]. While previous studies have explored the effects of anthropomorphic versus non-anthropomorphic images on attractiveness, there is a gap in addressing the comparative impact of dynamic and static presentations. Considering the prevalent use of dynamic visual content in modern media interfaces this study aims to investigate the influence of dynamic and static presentations on user emotional perceptions in autonomous vehicle displays.

In summary, anthropomorphism plays a significant role in products and services, yet its impact requires a multifaceted consideration, such as media presentation styles. With a thorough investigation, the study's objective is to offer specific design recommendations for future applications.

1.3 Research Questions

Currently, there is limited literature on integrating anthropomorphic images into car display panels, particularly within the transportation sector. Given the onboard panels in automated taxis as a pivotal point of customer interaction, understanding how the styles of these displays influence customer sentiment becomes imperative. Thus, this study aims to utilize eye-tracking technology to evaluate users' emotional responses to various images, providing valuable insights into the field. This research will explore the following two key questions:

- **Question 1:** How do dynamic and static anthropomorphic images in autonomous vehicle displays influence users' emotional responses?
- **Question 2:** Do anthropomorphic images in autonomous vehicle displays draw more attention than non-anthropomorphic ones, and how does this attention impact the ride experience?

2 Methods

2.1 Study Design and Participants

To explore passengers' emotional responses and visual attention towards autonomous vehicle displays, four distinct display designs were developed. We conducted an experimental test using eye-tracking technology to assess the emotional experiences of passengers during the ride. This study employs a 2×2 factorial design, with two primary factors: media presentation styles (static, dynamic) and anthropomorphism (non-anthropomorphic, anthropomorphic). This approach resulted in four display samples: (1) Static Non-Anthropomorphic (SN), (2) Static Anthropomorphic (SA), (3) Dynamic Non-Anthropomorphic (DN), and (4) Dynamic Anthropomorphic (DA). The anthropomorphic images are characterized by human-like features such as eyes and a mouth, whereas non-anthropomorphic images do not have these human facial characteristics (see Fig. 1). The displays are simulated in the current environment of Waymo's autonomous taxis.

Participants were voluntarily recruited from university users, aged 18 and above without eye conditions such as strabismus or nystagmus. The final experiment included 33 participants (13 Male, 20 Female), with an average age of 24 years (SD = 3.02). All participants had a normal or corrected-to-normal vision.

Fig. 1. Non-anthropomorphic (left) and anthropomorphic images (right).

2.2 Experimental Procedure and Data Collection

This study utilized the Tobii Pro Nano eye-tracking system, capturing gaze data at 60 Hz, with analysis performed using Tobii Pro Lab software (version 1.232). Stimuli were presented on a 1920 × 1080 pixels LCD screen with a 60 Hz refresh rate. Participants underwent a 9-point calibration process at the start, and each experimental stimulus began with a fixation cross to maintain participant attention. The study was conducted following Institutional Review Board (IRB) approval.

The study employed a repeated measures design, with each participant randomly viewing four different samples to mitigate potential biases, and the sample presentation order was counterbalanced across participants.

Upon arrival, participants were briefed on the experimental procedure and assured that they could withdraw from the experiment at any point without consequences, especially if they experienced any physical or mental discomfort. The experimental protocol began with a 9-point calibration, followed by participants watching a video illustrating a scenario of renting an autonomous taxi. Subsequently, participants viewed sample images and completed the User Experience Questionnaire (UEQ-S) and the Self-Assessment Manikin (SAM) scale. The first round of the experiment concluded at this stage. Each participant underwent four rounds of the identical experiment, and the entire experimental session was expected to last approximately 10 min. An overview of the experimental procedure is presented in Fig. 2.

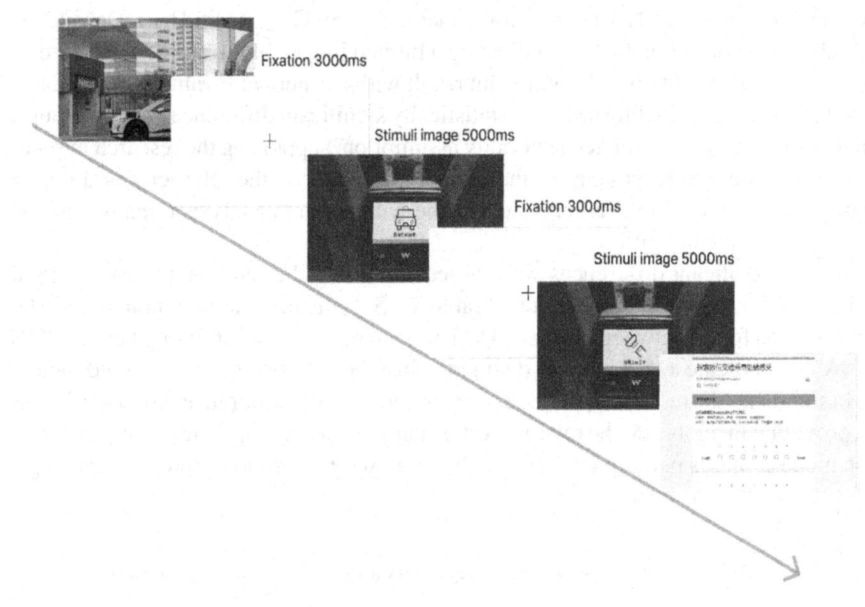

Fig. 2. Experimental process design.

3 Results and Analysis

Eye-tracking data, SAM scores, and UEQ-S scores from 33 participants were analyzed using IBM SPSS 22, focusing on average fixation duration and fixation count. Repeated Measures ANOVA was employed for analysis.

3.1 Average Fixation Duration

Table 1 presents the average fixation duration for each sample. The ranking of average fixation durations, from highest to lowest, is as follows: Dynamic Anthropomorphic Images, followed by Dynamic Non-Anthropomorphic Images, then Static Anthropomorphic Images, and finally Static Non-Anthropomorphic Images.

Table 1. Average Fixation Duration of Samples.

Sample	Dynamic Anthropomorphic (DA)	Dynamic Non-Anthropomorphic (DN)	Static Anthropomorphic (SA)	Static Non-Anthropomorphic (SN)
Average Fixation Time (s)	3.21 (SD = 4.45)	1.78 (SD = 1.39)	0.96 (SD = 0.3)	0.74 (SD = 0.23)

Statistical analysis using Repeated Measures ANOVA showed a sphericity assumption violation (Mauchly's $W = 0.006$, $p < 0.05$), addressed with Greenhouse-Geisser

(G-G) and Huynh-Feldt (H-F) corrections (factors: 0.389 G-G, 0.395 H-F). Despite the relatively small correction factors, indicating a limited impact of the corrections, within-subject effects testing showed a significant result with a spherical F value of 7.682 and a p-value of 0.000. This finding suggests statistically significant differences within groups, even in the presence of a violated sphericity assumption, supporting the research hypothesis of significant differences among the four samples. Despite the sphericity assumption violation, the corrected results indicate that the within-group analysis remains reliable and statistically meaningful.

Finally, significant differences were observed among the four sample groups (SN, SA, DN, DA) in pairwise comparisons (Table 2). Specifically, the only non-significant difference was found between DN and DA. This suggests that the similarity between DN and DA may indicate a lack of clear distinction between the two in the observed indicators. In summary, the results of pairwise comparisons support differences among the samples, providing insights into the relative performance of each group. The dynamic presentation mode captures participants' visual attention, with anthropomorphism attracting a modest amount of visual attention.

Table 2. Pairwise comparisons of the average fixation duration among the samples.

(I) Sample	(J) Sample	Mean Difference (I-J)	Std. Error	Sig.
SN	SA	−.223*	.061	.001
	DN	−1.039*	.242	.000
	DA	−2.477*	.775	.003
SA	SN	.223*	.061	.001
	DN	−.817*	.239	.002
	DA	−2.255*	.753	.005
DN	SN	1.039*	.242	.000
	SA	.817*	.239	.002
	DA	−1.438	.822	.090
DA	SN	2.477*	.775	.003
	SA	2.255*	.753	.005
	DN	1.438	.822	.090

3.2 Fixation Count

Table 3 shows the average fixation count for each sample. The ranking of average fixation counts, from lowest to highest, is as follows: Dynamic Anthropomorphic Images, followed by Dynamic Non-Anthropomorphic Images, then Static Anthropomorphic Images, and finally Static Non-Anthropomorphic Images.

Using Repeated Measures ANOVA for statistical analysis of sample fixation counts, the result of the sphericity test indicates the acceptance of the sphericity assumption (Mauchly's $W = 0.916$, $p > 0.05$). This suggests that the variance structure of the observed values is similar across groups. In within-subject effects testing, the assumed

Table 3. Sample Fixation Count.

Sample	Dynamic Anthropomorphic (DA)	Dynamic Non-Anthropomorphic (DN)	Static Anthropomorphic (SA)	Static Non-Anthropomorphic (SN)
Fixation Count	11 (SD = 4.75)	14 (SD = 4.81)	14 (SD = 3.9)	15 (SD = 5.13)

spherical F value is 5.254, with a significance of 0.002, indicating statistically significant differences within groups. Therefore, we can infer significant differences among the four samples in this indicator.

In conclusion, the pairwise comparisons of the results show statistically significant differences ($p < 0.05$) in fixation counts between SN and DA, SA, and DA, and DN and DA (Table 4). This highlights the distinctiveness in media presentation and anthropomorphism concerning users' visual search. The lower fixation counts in dynamic anthropomorphic images suggest a more focused visual attention.

Table 4. Pairwise comparisons of the Fixation Count.

(I) Sample	(J) Sample	Mean Difference (I-J)	Std. Error	Sig.
SN	SA	.576	1.031	.580
	DN	.606	1.083	.580
	DA	3.758*	1.076	.001
SA	SN	−.576	1.031	.580
	DN	.030	.937	.974
	DA	3.182*	1.049	.005
DN	SN	−.606	1.083	.580
	SA	−.030	.937	.974
	DA	3.152*	1.126	.009
DA	SN	−3.758*	1.076	.001
	SA	−3.182*	1.049	.005
	DN	−3.152*	1.126	.009

3.3 Self-assessment Manikin (SAM) Score

In the SAM scale, a 9-point Likert scale was used. When Valence approaches 9, it indicates that the participant perceives the emotion as highly pleasant. Similarly, when Arousal approaches 9, it signifies that the perceived emotion has a high level of excitement. When Dominance approaches 9, it suggests that the participant perceives the emotion with greater fluctuation.

From Table 5, it is evident that the ranking of Valence in terms of pleasantness is as follows: Dynamic Non-Anthropomorphic Images > Dynamic Anthropomorphic Images > Static Non-Anthropomorphic Images > Static Anthropomorphic Images. Similarly, the ranking of Arousal in terms of excitement is Dynamic Non-Anthropomorphic Images > Dynamic Anthropomorphic Images > Static Anthropomorphic Images > Static Non-Anthropomorphic Images. Finally, the ranking of Dominance in emotional fluctuation is Dynamic Anthropomorphic Images > Dynamic Non-Anthropomorphic Images > Static Anthropomorphic Images > Static Non-Anthropomorphic Images. Based on the above scores, it can be concluded that dynamic images elicit better emotional responses.

Table 5. SAM scale score.

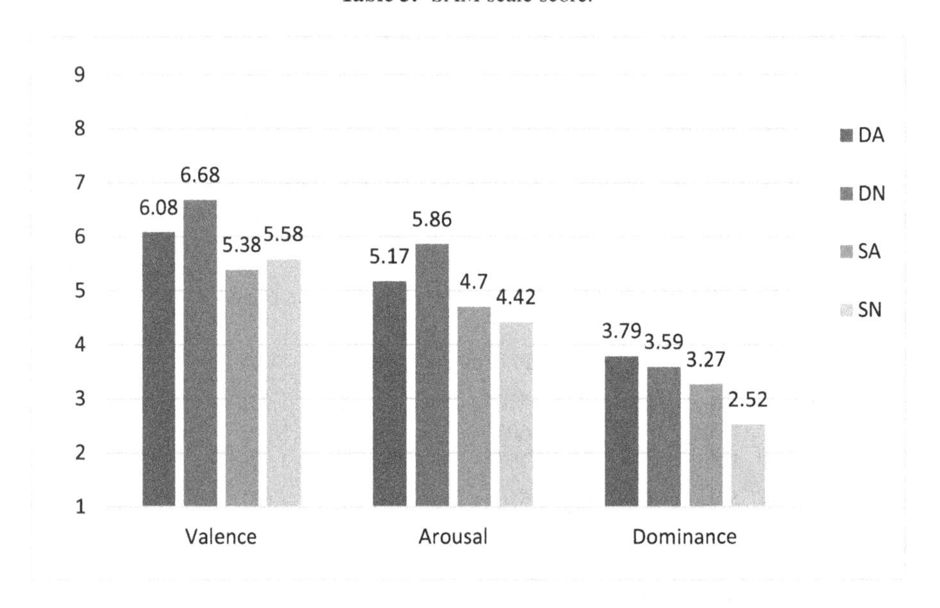

The study investigated the impact of emotional effects on different samples using the Self-Assessment Manikin (SAM). The study's statistical analysis, using Repeated Measures ANOVA, found a sphericity assumption violation (Mauchly's W = 0.635, $p < 0.05$). Corrected with Greenhouse-Geisser (0.733) and Huynh-Feldt (0.756) methods, significant differences were found in sample emotional responses.

Table 6 multiple comparison results indicate significant differences among samples in terms of emotions. Specifically, in the contrasts between samples, we observed significant differences in emotional performance between DN and both SA and SN. From the Valence and Arousal scores, it can be inferred that DN elicits significantly higher levels of pleasure and excitement compared to other samples. DA, on the other hand, exhibits the second-highest emotional effect, suggesting that dynamic images may evoke richer emotional experiences in individuals.

3.4 User Experience Questionnaire (UEQ-S) Scores

To understand the user experience perceptions of different samples, we utilized a 7-point Likert scale to collect participant experience data. Statistical analysis was performed

Table 6. SAM multiple comparison results.

(I) Sample	(J) Sample	Mean Difference (I-J)	Std. Error	Sig.
SN	DA	−.838*	.325	.011
	DN	−1.207*	.325	.000
	SA	−.278	.325	.395
SA	DA	−.561	.325	.087
	DN	−.929*	.325	.005
	SN	.278	.325	.395
DN	DA	.369	.325	.259
	SA	.929*	.325	.005
	SN	1.207*	.325	.000
DA	DN	−.369	.325	.259
	SA	.561	.325	.087
	SN	.838*	.325	.011

using Repeated Measures ANOVA, and the result of Mauchly's sphericity test indicated a violation of the sphericity assumption (Mauchly's $W = 0.033$, $p < 0.05$). To address this issue, we applied the Greenhouse-Geisser (G-G) and Huynh-Feldt (H-F) correction methods, obtaining correction factors of 0.397 (G-G) and 0.416 (H-F), respectively.

Table 7. User experience scale score.

Sample	Dynamic Anthropomorphic (DA)	Dynamic Non-Anthropomorphic (DN)	Static Anthropomorphic (SA)	Static Non-Anthropomorphic (SN)
Support	5.45(SD = 1.30)	5.79(SD = 0.85)	4.52(SD = 1.48)	5.18(SD = 1.15)
Simple	5.39(SD = 1.45)	6.33(SD = 0.78)	4.88(SD = 1.51)	6.39(SD = 0.70)
Efficiency	5.42(SD = 1.37)	5.94(SD = 1.17)	4.82(SD = 1.68)	5.58(SD = 1.30)
Clear	5.47(SD = 1.30)	6.45(SD = 0.56)	5.03(SD = 1.67)	6.21(SD = 0.86)
Exciting	4.64(SD = 1.56)	4.97(SD = 1.43)	3.82(SD = 1.42)	3.48(SD = 1.60)
Interesting	4.97(SD = 1.63)	5.3(SD = 1.57)	4.15(SD = 1.64)	3.27(SD = 1.52)
Innovative	4.70(SD = 1.46)	5.21(SD = 1.29)	4.42(SD = 1.37)	3.48(SD = 1.85)
Leading	4.30(SD = 1.57)	4.73(SD = 1.56)	3.82(SD = 1.35)	3.58(SD = 1.89)

In the main effects analysis results, we identified statistically significant differences in experiential data across various samples. More specifically, our observations indicate significant variations in the analysis when comparing DN to the other three samples. Additionally, we noted significant analytical distinctions between DN and SN. These results strongly imply diverse experiential perceptions among the different samples, underscoring the substantial impact of dynamic presentation methods on user experience.

Upon further analysis of the data in Table 7, we observed that the user experience scores for Dynamic Non-Anthropomorphic are significantly higher than those for other samples, followed by Dynamic Anthropomorphic. From this, we can infer that participants perceive dynamic presentation as providing a better user experience during riding.

4 Discussion

This study aims to investigate the impact of media presentation style and anthropomorphism on user experience in the context of in-vehicle displays for autonomous taxis. The following sections discuss eye movement patterns and emotional responses among users exposed to four display samples.

4.1 Impact of Media Presentation Style on Riding Experience

Through the analysis of eye-tracking data, we observed that the dynamic presentation mode received more gaze time and exhibited more positive fluctuations in emotional responses. These observations indicate that the dynamic presentation mode not only enhances the positive emotional experience but also contributes to increased support and overall riding experience. In terms of statistical data, we found a significant difference in gaze time between the dynamic presentation mode and static samples, indirectly suggesting a statistically significant difference in media presentation style. This result aligns with previous research, indicating that dynamic presentation style, compared to static images, elicits stronger emotional resonance from users, thereby enhancing the overall experience [15, 16]. Finally, when asked about their preferences for display presentation style, the results showed that 90.91% of participants favored the dynamic presentation mode.

4.2 Impact of Anthropomorphic Presentation

This study confirmed that the inclusion of anthropomorphic expressions in the samples successfully captured more attention and extended the average fixation duration. Evaluation using the SAM scale revealed that adding anthropomorphic expressions increased arousal levels and emotional fluctuations. However, it is noteworthy that the positive or negative emotional impact of anthropomorphic expressions has not been definitively determined. Although there is a correlation between the degree of anthropomorphism and emotional fluctuations, it does not always correspond directly to levels of liking and support. At the end of the experiment, participant preferences for anthropomorphic presentation were surveyed, and the results showed that 84.85% of participants leaned towards non-anthropomorphic displays. Future research could delve into the specific impact of different anthropomorphic expressions on emotional experiences.

4.3 Limitations

This study's limitations include potential biases due to the specific styles of samples used. Although intended to provide a comprehensive view of the riding experience, unexplored variations in design styles may have distinct effects on users' emotional responses.

Furthermore, the study limited the exploration of different emotional expressions of anthropomorphism and focused on a predominantly young demographic, so the findings may not broadly apply to other age groups.

5 Conclusions

This study explores the impact of media presentation and anthropomorphism on the user experience of an in-vehicle display for autonomous taxis, utilizing four display samples. A comprehensive evaluation, including eye-tracking data, SAM scales, and UEQ-S scales, reveals that participants exhibited heightened visual attention and enriched emotional experiences when exposed to dynamic images. Specifically, dynamic non-anthropomorphic images outperformed others in terms of average fixation time and visual search frequency, with dynamic anthropomorphic images ranking second. However, in the analysis of SAM scales, anthropomorphized images, despite inducing emotional fluctuations and excitement, did not exhibit a positive correlation with feelings of pleasure.

The research underscores the significant influence of media presentation and anthropomorphism on eye-tracking data and emotional experiences. Dynamic non-anthropomorphic images excel in visual guidance and emotional stimulation, while anthropomorphic elements notably contribute to increased fixation time. Future display style designs are advised to incorporate dynamic images for enhanced emotional experiences. Simultaneously, a thoughtful addition of anthropomorphic elements is recommended in scenarios where attracting visual attention is essential. Subsequent studies could explore detailed emotion analysis methods and conduct research in experimental environments closely resembling real-world ride experiences to deepen our understanding of the effects of anthropomorphism in the ride experience.

Acknowledgments. This study was funded by the National Science and Technology Council (grant number 112–2221-E-027–080-MY2).

Disclosure of Interests. The authors have no competing interests to declare that are relevant to the content of this article.

References

1. Stocker, A., Shaheen, S.: Shared automated vehicles: review of business models. International Transport Forum Discussion Paper No. 2017–09 (2017)
2. Nutonomy launches. https://reurl.cc/1304kY. Accessed 05 Jan 2024
3. BBC News Chinese. https://reurl.cc/krq09d. Accessed 05 Jan 2024
4. California Public Utilities Commission. https://reurl.cc/805n7j. Accessed 06 Jan 2024
5. Shen, W., Lopes, C.: Managing autonomous mobility on demand systems for better passenger experience. In: Chen, Q., Torroni, P., Villata, S., Hsu, J., Omicini, A. (eds.) PRIMA 2015. LNCS (LNAI), vol. 9387, pp. 20–35. Springer, Cham (2015). https://doi.org/10.1007/978-3-319-25524-8_2

6. Carreyre, F., Coulombel, N., Berrada, J., Bouillaut, L.: Economic evaluation of autonomous passenger transportation services: a systematic review and meta-analysis of simulation studies. Rev. Econ. Ind. **178–179**, 89–138 (2022)
7. Muralidhar, P., Sai Prashanth, A., Pavan Kumar, K., Rani, C., Rajesh Kumar, M.: Accident prevention for autonomous vehicles. In: 2023 2nd International Conference on Vision Towards Emerging Trends in Communication and Networking Technologies (ViTECoN), IEEE Xplore (2023)
8. Freedman, I.G., Kim, E., Muennig, P.A.: Autonomous vehicles are cost-effective when used as taxis. Inj. Epidemiol. **5**, 24 (2018)
9. Hur, J.D., Koo, M., Hofmann, W.: When temptations come alive: how anthropomorphism undermines self-control. J. Consum. Res. **42**(2), 340–358 (2015)
10. Zhang, Y., Cao, Y., Proctor, R.W., Liu, Y.: Emotional experiences of service robots' anthropomorphic appearance. Ergonomics **66**(12), 2039–2057 (2023)
11. Cao, Y., Proctor, R.W., Ding, Y., Duffy, V.G., Zhang, Y., Zhang, X.: Is an anthropomorphic app icon more attractive than a non-anthropomorphic one a case study using multimodal measurement. Int. J. Mob. Commun. **20**(4), 419 (2022). https://doi.org/10.1504/IJMC.2022.123789
12. Chiang, A.-H., Trimi, S., Lo, Y.-J.: Emotion and service quality of anthropomorphic robots. Technol. Forecast. Soc. Chang. **177**, 121550 (2022)
13. Martina, M., Appel, M., Gnambs, T.: Human-like robots and the uncanny valley. Zeitschrift Fur Psychologie-J. Psychol. **230**(1), 33–46 (2022)
14. Pei, H., Huang, X., Ding, M.: Image visualization: dynamic and static images generate users' visual cognitive experience using eye-tracking technology. Displays **73**, 102175 (2022)
15. Wang, C.-H., Shih, Y.-H., Lo, Y.-C., Huang, C.-Y.: Researching the adolescent visual color perception and brain wave in dynamic images. In: 2016 International Conference on Advanced Materials for Science and Engineering (ICAMSE), pp. 162–164. (2016)
16. Sato, W., Yoshikawa, S.: Emotional elicitation by dynamic facial expressions. In: Proceedings of the 4th International Conference on Development and Learning, pp. 170–174 (2005)

Analyzing the Remote Operation Task to Support Highly Automated Vehicles – Suggesting the Core Task Analysis to Ensure the Human-Centered Design of the Remote Operation Station

Hanna Koskinen[1](✉), Andreas Schrank[2], Esko Lehtonen[1], and Michael Oehl[2]

[1] VTT Technical Research Centre of Finland Ltd., VTT, Box 1000, FI-02044 Espoo, Finland
{hanna.Koskinen,esko.Lehtonen}@vtt.fi

[2] German Aerospace Center (DLR), Institute of Transportation Systems, Lilienthalplatz 7, 38108 Braunschweig, Germany
{andreas.schrank,michael.oehl}@dlr.de

Abstract. Intelligent technology and high automation solutions will need to be integrated into the future's sustainable traffic systems. However, for a long time, automatic vehicles cannot be expected to cope with all possible traffic situations without a human intervention. In the EU-funded Hi-Drive project, remote operation has been proposed as one feasible solution for these situations. However, the remote operation of an automatic vehicle is a new work task/profession whose requirements are not yet properly known. In this paper, we propose the core task analysis as a way to better comprehend the challenges of the safety-critical task of remotely controlling highly automated vehicles in open road settings and demonstrate the core task analysis method with the operative situations from the scenario catalogue. The operating environment imposes certain demands on the operators, who need to use their trained skills and knowledge to deal with them. The results of the core task analysis highlight the environmental demands and the critical resources and competence the human operator should have. Core task analysis may be most helpful in comprehending the demands of remote operator tasks in supporting the highly automated vehicles as well as carrying out human-centered design of operating workplace.

Keywords: Automated Vehicle · Remote Operation · Remote Assistance · Remote Driving · Human-Centered Design · Task Analysis

1 Introduction

Highly automated vehicles (SAE Level 4 to 5) [1] can improve the sustainability of the transport system in the future, for example, by improving mobility and safety of the traffic system. EU H2020 funded Hi-Drive project aims to improve the operational

H. Krömker (Ed.): HCII 2024, LNCS 14732, pp. 145–156, 2024.
https://doi.org/10.1007/978-3-031-60477-5_11

design domains (ODD) and minimize the need for human intervention by developing advanced technologies that enable high-performance connected automated driving functions (CADFs). The project will test and demonstrate concepts of CADFs for passenger cars and professional transport (e.g., trucks) in various traffic situations on highways and urban areas. Despite the recent advances in automated driving technology, automated vehicles are still not able to manage on many situations on open roads. Human remote operation has been recognized by the Hi-Drive project as a solution, which would enable operating highly automated vehicles (HAVs) without having a human capable taking over the driving onboard the vehicle. Remote operation of HAVs, however, is a new field of profession and thus may come with many new challenges and user needs that can be significantly different from designing an automated driving system. Furthermore, the implementation of remote control in automated driving also requires the design of the remote operation workplace and the associated human-machine interface (HMI) and human-computer interaction (HCI).

1.1 Remote Operation of Automated Vehicles

Highly automated driving is envisioned to increase safety and efficiency, shorten travel duration, and reduce traffic congestion while emitting less greenhouse gases than driving without an automation [2]. A prerequisite for the benefits of automation is that they can function well in the normal traffic. However, even on a high level of automation (SAE Level 4), situations may emerge that the HAV cannot resolve autonomously, and human intervention is needed. Thus, to continue the ride safely, efficiently, and user-friendly, the involvement of a human remote operator has been proposed. Instead of being located aboard of the HAV, in remote operation, the human operator could be in a control center and remotely review and resolve issues that the driving automation cannot handle by itself [3, 4]. Different approaches of remote operation are conceivable: In "remote driving", the remote operator is responsible for executing direct longitudinal and lateral control of the HAV, that is, the dynamic operational driving task, by steering, accelerating, and braking. In contrast, remote operation mode of "remote assistance" does not expect the remote operator to execute the dynamic driving task but to provide high-level guidance and support on how to resolve a situation with a HAV system error or exceeding system limit [5, 6].

Regardless of which mode of remote operation (i.e., remote driving or remote assistance) is in question, the remote operation of HAVs can be considered a safety-critical activity. In order to provide a sound concept of operations and interaction between remote operator and HAV and the subsequent design of a HMI for safe and efficient remote operation, specific situations shall be considered in which an HAV requires assistance or even a complete takeover of the dynamic driving task. Therefore, a thorough specification of possible use cases and scenarios for the introduced variants of remote operation is helpful. Even though system limits can be specified as "operational design domains" (ODDs), there is a plethora of situations that cannot be represented using ODDs. Thus, there are attempts in the literature to compile potential use cases and scenarios for the operation of HAVs in which involvement of a human remote operator would be postulated [7].

1.2 Human-Centered Design

Human-centered design (HCD) is a design approach that focuses on users and their needs, behaviors, and experiences. It aims to deeply understand the users and create design solutions that address the specific challenges and desires of users. The ISO 9241–210 [8] standard defines HCD as "*an approach to systems design and development that aims to make interactive systems more usable by focusing on the use of the system and applying human factors/ergonomics and usability knowledge and techniques.*"

The HCD process begins with creating an understanding and specifying the context of use. The first phase of the process is an essential and determines the later ones, that is, eliciting relevant user requirements, producing appropriate design solutions, and evaluating the solutions with integrating the user feedback and experiences. Thus, the HCD is iterative and flexible in the sense that the process allows learning from the user feedback and improvements of the solutions over time. The HCD process and involvement of user knowledge is motivated and expected to result design solutions that are more innovative, intuitive to use and accessible for users as well as generally better meet the expectations and preferences of the users [9]. Moreover, HCD have suggested to improve customer satisfaction, loyalty, and profitability of the design solution.

The user perspective provided by the HCD may also contribute greatly to the design of automatic transportation systems [10, 11]. In the particular context of remote operation of HAVs, HCD principles have guided the design of a workplace for remote operation. First, the context of use was analyzed and specified by viewing footage of HAV operations in the real world, observing employees of control centers in public transport and interviewing them about their work. User requirements for future control centers in public transport were compiled based on a task analysis with control center employees [12]. The above-mentioned analysis was based on a similar remote operation work in the transportation context, however, not specifically on the remote operation of HAVs that is important to acknowledge in drawing conclusions for any specific user needs and criteria. This compromise was accepted as the actual context of use, the remote operation of HAVs on open roads, did not exist at this point and could therefore not be analyzed. Hence, based on the existing sources of information on similar operating tasks, a list with initial design requirements was compiled. Next, key requirements were addressed in a preliminary design solution, a wireframe of a remote operation HMI. The wireframe was then elaborated until a paper-and-pencil prototype emerged. According to the iterative process of HCD, the early-phase prototype was further developed to the stage of a click prototype that was then evaluated with the representative users. Consequently, applying a comprehensive usability test, the click prototype was evaluated with traffic control center staff [3]. Considering the quantitative results of this usability test and the participants' qualitative feedback, the click prototype was reiterated to a prototypical workplace. Subsequently, this workplace was tested in an experimental study in a laboratory setting with participants that fulfilled the educational requirements for a Technical Supervisor, the legal definition of a remote assistant in Germany [13, 14]. Lastly, these results may inform an even more in-depth development of the remote operation workplace in the future and its continuous adjustment to a potentially dynamic context of use.

Many user-driven design approaches and techniques have been introduced to serve the HCD. All these methods have developed different ways to generate contextual understanding and help users to envision "use before actual use" and, in doing so, enable users to state their opinions and perspectives regarding the proposed design solution. For example, one way to explore the potential impact of new design concepts on users' experiences is to use scenario-based methods that illustrate how the product would be used in various usage situations [15, 16]. Scenarios are narratives that depict the product in realistic and yet specific situations, enabling users to envision the advantages as well as disadvantages of the proposed solution through storytelling and description. Another much used method in HCD is the use of different types of prototypes as demonstrated in the above example of designing remote operation of HAVs to foster the discussion about the proposed solutions [17]. By grounding the discussions in concrete artifacts, prototypes can help to clarify the design goals, assumptions, and constraints that the users may see in the solution and in that way enable a gradual modification of the solution towards an appropriate result.

However, only concentrating on task- and situation-specific solutions, it is difficult to master the so-called task-artifact cycle that refers to the co-evolving nature of the new design solution and the activity it is supposed to support [18]. According to this cycle, the activity shapes the design requirements and, when the design solution is implemented, it changes the activity itself. This creates a (task-artifact cycle), where the design and the activity are never in a stable state and the activity may evolve in unexpected ways. To deal with this continuous mismatch of eliciting proper design requirements and creating solutions to support the activity, and to enable the formative and intentional design of tools and their future use, future needs must be understood and comprehended in more general terms.

In this paper, we introduce a core task analysis (CTA) [19] as a methodical mean to understand the generic demands of the remote operation of HAVs and to be able to develop the remote operator work and related tools (e.g., remote operation workplace and HMI) with long-term overall goals in mind. As the remote operation of HAVs is a new and emerging field of operation in the context of road transportation, CTA may be most helpful in comprehending the work demands of remote operator tasks in supporting the HAVs as well as conducting HCD process to develop appropriate tools. CTA may also contribute to developing the automated vehicle industry more extensively as it provides a rich description of the whole system within which HAVs are operated.

The continuation of this paper is structured as follows: First, the core-task method and the data used for demonstrating its application in the context of the remote operation of automated vehicles are introduced. Second, results are presented with examples demonstrating the insights that may be achieved with CTA in studying remote operation in context of HAVs. Our paper concludes with a discussion on how CTA can facilitate/enhance the HCD process and inclusion of in-depth user knowledge and contextual understanding, and how CTA approach relates to the broader context of designing highly automated transport systems.

2 Method and Materials

The core-task analysis (CTA) is suggested as a method to understand the requirements of the remote operation task and thus to ensure a well-grounded and human-centered design of the remote operation station.

2.1 A Core Task Analysis

A core task analysis (CTA) is a method for studying the complex and often highly automated and technologically advanced work activity in various industrial domains, such as the work of process control operators [19, 20]. The CTA aims to understand how and why the human actor (e.g., the remote operator) and the environment, including organizational structures and tools (e.g., the remote operation center with different workstations), coordinate their actions to achieve the goals and purposes of the work activity (e.g., a safe operation of automated vehicles and generally the safety of the traffic system). The resulting understanding of the functional content of the work may help set the relevant user requirements for developing the work practices and tools. With the help of CTA, for example, the work demands for remote operation of automatic container cranes [21], the change in the concept of operations and metro drivers' role related to the introduction of an automatic metro service [10] and in maritime piloting [22] have been clarified.

The core task can be defined as "*the generic key content that is relatively stable and unaffected by the current ways of accomplishing the activity, toward which core the activity should be oriented in order to be appropriate and develop*" [20]. Consequently, the CTA differs from many other task analysis methods because it does not prescribe how the actors should perform their actions (sequential order of individual tasks) in the work domain but instead, it reveals the existing possibilities that the actor can use to achieve their goals and cope with the challenges they face. It can be seen to provide a map of the work domain with all the possible pathways the operation can take, rather than a predefined and fixed route that should be taken [23]. Thus, the CTA helps to understand the object of activity and its demands from the human actors' point of view in a more general level that is particularly essential and helpful when designing solutions for a new work activity such as remote operation of HAVs.

The work activity under scrutiny is analyzed by means of a model that identifies the core task of an activity or work domain based on three inherent aspects of operational environment, that are, the dynamicity of the environment (e.g., temporal demands, such as timing, duration, and delays), the complexity of the relations among its elements, and the uncertainty of the environment (e.g., due to our limited knowledge). Moreover, the CTA considers three kinds of resources that the human actor may exploit, that are, skill-, knowledge-, and collaboration-related human capabilities. By combining the three aspects of the operational environment and the human resources, it is possible to obtain a three-by-three matrix (often also illustrated a start-shaped figure see Fig. 1) that defines the nine core tasks (e.g., skills to deal with dynamicity, complexity, and uncertainty) of the work activity under scrutiny.

The specific content of core tasks is derived from modelling the work activity. Core tasks must be specified for each activity and domain separately as they are contextually grounded [20]. Typically, the core tasks reflect some inherent contradictions of the object

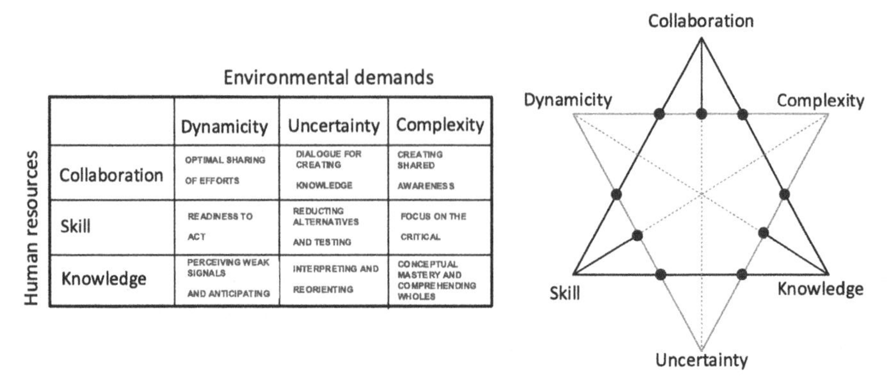

Environmental demands / Human resources

	Dynamicity	Uncertainty	Complexity
Collaboration	OPTIMAL SHARING OF EFFORTS	DIALOGUE FOR CREATING KNOWLEDGE	CREATING SHARED AWARENESS
Skill	READINESS TO ACT	REDUCTING ALTERNATIVES AND TESTING	FOCUS ON THE CRITICAL
Knowledge	PERCEIVING WEAK SIGNALS AND ANTICIPATING	INTERPRETING AND REORIENTING	CONCEPTUAL MASTERY AND COMPREHENDING WHOLES

Fig. 1. A CTA framework in which the work activity is considered from its domain specific environmental demands and human resources point of views to comprehend the contextually grounded core task functions.

that need to be resolved in activity. Furthermore, a core task is not a fixed or static concept by its nature, but rather a dynamic and evolving one that reflects the changing nature of work and its context. Consequently, the core task can be used as a framework to guide the design, evaluation and improvement of work systems, processes, and outcomes. Wahlström et al. [24] provide an example of applying the CTA in designing ship-bridge solutions in maritime transportation context.

2.2 Data

A scenario catalogue for the remote operation of HAVs in public transport [7] was used as a main source of data collection to demonstrate and conduct the CTA to test its methodical feasibility in analyzing the remote operation of automated vehicles. The scenario catalogue includes 74 core scenarios related to the operational situations that may be realized in remote operation of HAVs.

Various sources were tapped for the scenarios. They were collected in research projects on automated vehicles that the authors were involved in. These include real-world laboratories of highly automated shuttles in public transport. The scenarios originated from interviews with and observations of control center staff in Hamburg, Germany, video analyses from naturalistic road events in the EU project CityMobil2, in La Rochelle, France, and Trikala, Greece, and interviews with on-board operators of AVs.

The scenarios are structured by the mode of remote operation, that is, remote driving and remote assistance, and by the actors involved: the remote operator (RO), automated vehicle (AV), passengers (P), the infrastructure (I), and other actors (OA). The focus of the catalogue is on the remote operator tasks performed in the operation mode of remote assistance, but remote driving is also considered. Each scenario is categorized in a use case cluster that depends on the agents of the respective interaction as well as a use case. The structure of each scenario follows the chain of cause, event, and consequence and can be amended with the measures required to resolve the issue. It has to be noted that the scenario catalogue is a "living document" in the sense that it is not exhaustive. As technology and use cases of HAVs advance, new scenarios may emerge and will therefore need to be added to the catalogue. Likewise, new actors or interactions among them could become relevant over time.

3 Results and Discussion

Demonstration of CTA was performed based on the scenario catalogue depicting the remote operation of HAVs. The main vantage point to the scenario catalogue was the scenarios considered to be included in the tasks of the remote assistance.

3.1 Environmental Demands

According to the CTA, the uncertainty features play a significant role in remote operation of HAVs. These features include general uncertainty due to the lack of experience with remote assistance of HAVs (e.g., experiences gained from exceptional situations), obstacles such as other vehicles that may or may not pose a threat, passenger behavior that may conflict with instructions or rules and interaction challenges with vulnerable road users (VRUs) such as pedestrians or cyclists that may appear unexpectedly. Moreover, the remote assistance may be challenged by technical malfunctions such as sudden stops, fires, charging issues or braking problems, condition of video feeds or sensors that may be affected by dirt or weather. The disruptions in road infrastructure may impair the connection between HAV and infrastructure causing uncertainty due to the lack of operational information. Many operational situations involving a high degree of uncertainty are also such that the appearance and duration of the situations is difficult to predict. Therefore, it is essential to develop robust and adaptive methods for coping with uncertainty and ensuring resilience in the HAVs operation.

The complexity of the remote operation task may not be a remarkable challenge; however, much depends on the final operational concept and the design solution, for example, what operational situations and tasks are included in the remote operator's job description that provide remote assistance for HAVs. Whereas when considering remote operation from remote driving point of view the environmental demands that challenge the operation (i.e., dynamic driving task) would be different in many ways. The remote operation of a HAV in urban scenarios involves various technical and operational challenges related to complexity. The remote operator needs to be aware of the HAV's capabilities and limitations, as well as the dynamic and diverse traffic conditions on open roads. For example, the remote operator may have to intervene when the HAV encounters a novel situation, such as an unmapped construction site or a deadlock at an intersection. The remote operator may also have to communicate with other road users or authorities in case of an emergency or a collision situation. The complexity that the remote operator face is that unusual circumstances in the HAV operation are often different and thus any predefined procedures or task protocols for handling the situation can be difficult to provide. These tasks may require a high level of technical understanding, situation awareness, and decision-making skills from the operator.

According to the CTA, many dynamic features in the operating environment challenge the remote operators' work. Traffic is a dynamic concept, for example, it is referred to with various dynamic terms such as continuous traffic flow, changing speeds and traffic situations. Thus, also the remote operation of HAVs is much about collaboration and communication with the other road users in dynamic traffic situations. In some situations, the remote operator must interact with passengers and communicate safety-critical information to them in a timely manner. For instance, communication with passengers is a legal requirement for operating HAVs on public roads in Germany in the event of a minimum-risk maneuver [13]. Also, the remote operator must act quickly in emergencies, such as collisions or medical emergency. In such situations, time-critical actions may need to be performed to appropriately handle the situation. In the case of remote assistance, the passenger communication seems to be particularly of importance. One open design question potentially affecting much to the dynamicity of the HAV operations is the planned operating schedule (e.g., at what intervals the HAVs are expected to serve the passengers), a decision that partly defines the impact of delays due to the traffic disruptions and the time demands placed for the remote operation.

3.2 Core Task Functions

Core task functions are elicited from the analysis of the human actor's means for managing the domain/environmental characteristics related to dynamism, complexity, and uncertainty. To perform the remote operation effectively in different traffic situations, the operators need to cooperate, apply their skills, and use their knowledge. When each of the domain specific characteristics is examined through human resources as means to manage the work tasks, the general core-task functions of the work emerge. In Table 1, the core task functions of remote assistance are elicit based on the scenario catalogue.

Table 1. Summary table of core task function in remote assistance of HAVs.

		Environmental demands		
		Dynamicity	Uncertainty	Complexity
Human resources	Collaboration	Ability to function as a first point of contact with the passangers, delivery of (time critical) info/instr. Following through the area of own responsibility, may change depending on the RO assignment.	Sharing experiences with the colleagues in the CR on complex HAV remote operating situations. RO situation training, mental image training, and repetition.	Clear communication with other stakeholders to maintain SA and confirm safe operation. Combining competencies in problem solving and sharing responsibility with other professionals.
	Skill	Ability to react situations with very different dynamism (HAV assisting vr. traffic emergency handling). Ability to make decisions under time pressure (e.g., passenger communication) and in calm way.	The ability to work under uncertain information and mastery of strategies to test, reduce the amount of alternative solutions and anticipation (e.g., defining best pathway for HAV).	Prioritizing the remote assistance task order depending on the situational characteristics.
	Knowledge	Sensitivity for characteristics of different operational situations and ability to reflect those against the operation and its shedule.	Knowing whose should be involved in solving the specific operating situation. Knowledge on the specific open road operating environment of the HAVs.	Sufficient training and knowledge on the HAV's technical implementation and restrictions.

The introduction of HAVs on open roads may demand new human actor, that is, the remote operator to get involved. It is obvious that remote operation of a HAV is significantly different from driving a car. However, also within the framework of remote operation, whether the operation is in the mode of remote assistance or remote driving has implications to the core-tasks of the remote operators. Moreover, the initial inquiry of the scenario catalogue revealed that also the design decisions on the operational concept may affect the content of the core tasks exceedingly.

The remote operators need to acquire new skills and knowledge to remotely operate the HAVs (e.g., the used technologies and operating strategies), and to participate collaborative operation of the traffic system. The CTA of the remote assistance highlighted the importance of the collaboration in handling the anticipated operative situations in the scenario catalogue. In remote assistance, the remote operator is expected to be the first point of contact with the passengers and calmly convey them information and instructions in situations which from the passengers' point of view may be distressing and most often out of normal. Moreover, the communication and collaboration among the different actors (e.g., other road users, maintenance personnel, and officers) changes significantly when one key actor (i.e., the remote operator) in the situation participates remotely. Handling the situations demand well thought through operating procedures and clearly defined responsibilities between the parties in order to restore the normal traffic flow after a disruption. Especially in the early stages of the operational life cycle, sharing experiences with the colleagues on complex HAV operations is important to accumulate operating experiences and develop the profession and work processes.

There are skills that are important for remote assistance such as an ability to react situations with very different dynamics, for example, assist HAV to go around the road work site or handle emergency like accidents. In these situations, the remote operator may need to make decisions under time pressure and prioritize the order of tasks based on the specific characteristics of each individual operative situation. The remote operator also must tolerate certain amount of uncertainty as according to the CTA it plays a significant role in remote operation of HAVs. Thus, in remote assistance the operator must be able to work under uncertain information and master skills to check and test and to reduce alternatives on how to best ensure a safe and smooth operation. Operators also need to deal with possible new sources of uncertainty that may arise from the automatic system.

The complexity of remote assistance may not be a remarkable challenge; however, much depends on the final operational concept and the design solution of HAV (e.g., technical implementation and operative envelope) that content the remote operator must have sufficient training and knowledge on. In addition to the technical functioning of the system, the operator has to have knowledge on the specific open road environment that the HAVs operate so that it would be easier to comprehend and create an appropriate situation awareness about the traffic situation that the HAV is in.

3.3 Methodical Considerations on CTA in Analyzing HAV Operations

The study explored the feasibility of the CTA —originally developed for understanding industrial work activity in safety-critical domains such as nuclear power process control for automated driving. It was assumed that it may be beneficial to find methodological input for understanding human behavior regarding highly automated systems from other domains in which automation is extensively used. Thus, taking advantage of HCD approach used in domains with tradition in high automation solutions, the study was a conceptual exercise in which CTA was applied to examine remote operation of HAVs in open road transport.

Even though the material primarily used in the CTA is often based on empirical data from the studied work context (e.g., interviews and observation), in the analysis of remote operation of HAVs, we took advantage of the readily available descriptions of operational situations in the scenario catalogue. The scenarios provided a good basis for the CTA, however, the findings of this study can only provide preliminary information about the work demands of the remote operation and whether the CTA approach can be used for identifying relevant phenomena in HAVs operation. To exploit fully the benefits, CTA would require empirical inquires to be the main source of the information used in the analysed instead of a desktop-type study that was now conducted. Despite of this, the findings of CTA appear to be most useful regarding understanding the remote operation of HAVs. The CTA not only brings meaningful content to core-task functions, but also may present relevant design principles to guide the development of automated traffic systems. The CTA focused on the tactical and operational remote operation tasks, highlighting issues such as the meaning of communication/ collaboration, anticipation, and shared awareness in the remote operation of HAVs.

Based on this explorative study in the context of automated driving, we suggest three areas for the CTA to provide specific contributions (1) HCD of remote operation workplace and related HMI by eliciting user requirements and acceptance criteria for their adequate evaluations, (2) designing remote operator training and curriculum, and (3) investigating human operator behavior in empirical studies.

4 Conclusion

In the remote operation of HAVs is necessary to ensure that HAVs can be introduced to a wide range of traffic situations in a safe and efficient manner. This paper introduced CTA as a method to understand the requirements of the HAV remote operation task. CTA identified that uncertainty plays a significant role in the remote operation of HAVs. Uncertainty emerges from the problems with the connectivity between the remote operation and the HAVs but also from the difficulty to predict what will happen next. Another challenge for the remote operation is the dynamic nature of traffic which dictates that the remote operator must act in a timely manner to not obstruct the traffic flow and thus contribute to congestions. Remote assistance of stopped or slowly moving HAVs has lower dynamic requirements than remote driving. However, also a remote assistant may need to act quickly when communicating and assisting passengers in safety-critical situations. In contrast, the complexity of the situations may not be a remarkable challenge, especially for remote assistance but this depends on what tasks and solutions are included in the concept of operation. Based on our initial exploration of CTA in the context of remote operation of HAVs, the CTA not only brings meaningful content to the description of the core tasks but also may help in the elicitation of user requirements and acceptance criteria for their adequate evaluations in accordance with the HCD.

Acknowledgments. This project has received funding from the European Union's Horizon 2020 research and innovation programme under grant agreement No. 101006664. The author(s) would like to thank all partners within Hi-Drive for their cooperation and valuable contribution.

References

1. SAE: Surface vehicle recommended practice J3016: Taxonomy and Definitions for Terms Related to Driving Automation Systems for On-Road Motor Vehicles (2021). https://doi.org/10.4271/J3016_202104
2. Goldin, P.: 10 Advantages of Autonomous Vehicles. ITSdigest (2018). https://www.itsdigest.com/10-advantages-autonomous-vehicles
3. Kettwich, C., Schrank, A., Oehl, M.: Teleoperation of highly automated vehicles in public transport: user-centered design of a human-machine interface for remote-operation and its expert usability evaluation. Multimodal Technol. Interact. **5**(5), 26 (2021). https://doi.org/10.3390/mti5050026
4. Kettwich, C., Schrank, A.: Teleoperation of highly automated vehicles in public transport: state of the art and requirements for future remote-operation workstations. In: 27th ITS World Congress, Hamburg, Germany (2021)
5. Automated Vehicle Safety Consortium (AVSC). In: AVSC Best Practice for ADS Remote Assistance Use Case. (AVSC-I-04–2023). SAE Industry Technologies Consortia (2023)
6. Society of Automotive Engineers: Taxonomy and Definitions for Terms Related to Driving Automation Systems for On-Road Motor Vehicles. (SAE J 3016–202104). SAE, Washington, D.C. (2021). https://www.sae.org/standards/content/j3016_202104
7. Kettwich, C., Schrank, A., Avsar, H., Oehl, M.: A helping human hand: relevant scenarios for the remote operation of highly automated vehicles in public transport. Appl. Sci. **12**(9), 4350 (2022). https://doi.org/10.3390/app12094350
8. International Organization for Standardization. Geneva, Switzerland: International Organization for Standardization; ISO 9241–11:1998 Ergonomic requirements for office work with visual display terminals (VDTs) -- Part 11: Guidance on usability (1998)
9. Kujala, S.: Effective user involvement in product development by improving the analysis of user needs. Behav Inf. Technol **6**(27), 457–473 (2008)
10. Karvonen, H., Aaltonen, I., Wahlström, M., Salo, L., Savioja, P., Norros, L.: Hidden roles of the train driver: a challenge for metro automation. Interact. Comput.Comput. **4**(23), 289–298 (2011)
11. Koskinen, H., Aromaa, S., Goriachev, V.: Human factors engineering program development and user involvement in design of automatic tram. IADIS Int. J. Comput. Sci. Inf. Syst. **2**(16), 61–76 (2021)
12. Kettwich, C., Dreßler, A.: Requirements of future control centers in public transport. In: ACM (Chair), AutomotiveUI 2020: 12th International Conference on Automotive User Interfaces and Interactive Vehicular Applications, Washington, DC, USA (2020)
13. Gesetz zur Änderung des Straßenverkehrsgesetzes und des Pflichtversicherungsgesetzes - Gesetz zum autonomen Fahren, Bundesgesetzblatt (2021). https://www.bgbl.de/xaver/bgbl/start.xav?startbk=Bundesanzger_BGBl&start=//*[@attr_id=%27bgbl121s3108.pdf%27]#__bgbl__%2F%2F*%5B%40attr_id%3D%27bgbl121s3108.pdf%27%5D__1649730045177
14. Schrank, A., Walocha, F., Brandenburg, S., Oehl, M.: Human-Centered design and evaluation of a workplace for the remote assistance of highly automated vehicles. Cogn. Technol. Work (2024). https://arxiv.org/pdf/2308.02330

15. Rosson, M.B., Carroll, J.M.: Narrowing the gap between specification and implementation in object-oriented development. Scenario-Based Design Envisioning Work Technol. Syst. Dev. **247**, 278 (1995)
16. Nielsen, J.: Scenarios in discount usability engineering. In: Carroll, J.M. (ed.) Scenario-Based Design: Envisioning Work Technology in System Development, pp. 59–83. Wiley, NY (1995)
17. Bødker, S., Grønbæk, K.: Cooperative Prototyping: users and designers in mutual activity. Int. J. Man Mach. Stud. **34**, 453–478 (1991)
18. Carroll, J.M.: Designing Interaction: Psychology at the Human-Computer Interfac., Cambridge University Press (1991)
19. Norros, L.: Acting Under Uncertainty: The Core-Task Analysis in Ecological Study of Work. VTT (2004)
20. Norros, L. Savioja, P., Koskinen, H.: Core-Task Design: A Practice-Theory Approach to Human Factors. Synthesis Lectures on Human-Computer Informatics. Springer, Cham (2015). https://doi.org/10.1007/978-3-031-02211-1
21. Koskinen, H., Karvonen, H., Haggren, J.: Enhancing the user experience of the crane operator: comparing work demands in two operational settings. In: Proceedings of the 30st European Conference on Cognitive Ergonomics, pp. 37–40 (2012)
22. Nuutinen, M., Norros, L.: Core task analysis in accident investigation: analysis of maritime accidents in piloting situations. Cogn. Technol. Work. Technol. Work **11**(2), 129–150 (2009)
23. Norros, L.: Understanding acting in complex environments: building a synergy of cultural-historical activity theory, peirce, and ecofunctionalism. Mind Cult. Act. **25**(1), 68–85 (2018)
24. Yang, C.Y.D., Fisher, D.L.: Safety impacts and benefits of connected and automated vehicles: how real are they? J. Intell. Trans. Syst. **25**(2), 135–138 (2021). https://doi.org/10.1080/15472450.2021.1872143
25. Wahlström, M., Karvonen, H., Norros, L., Jokinen, J., Koskinen, H.: Radical innovation by theoretical abstraction - A challenge for the user-centred designer. Des. J. **19**(6), 1–21 (2016). https://doi.org/10.1080/14606925.2016.1216210

Design of a Virtual Assistant: Collect of User's Needs for Connected and Automated Vehicles

Julie Lang[1](✉), François Jouen[1], Charles Tijus[2], and Gérard Uzan[3]

[1] EA 4004 - CHArt, EPHE-PSL, Paris, France
julie.lang@ephe.psl.eu
[2] EA 4004 - CHArt-LUTIN, UPL - Université Paris 8, Saint-Denis, France
[3] EA 4004 - CHArt-THIM - Université Paris 8, Saint-Denis, France

Abstract. Intelligent virtual assistants are now present in our life, fulfilling tasks for helping us in many activities to ease our daily life. These digital technologies integrate diverse functionalities, communicating and cooperating with information systems, with the aim of being appealing to the widest possible range of users. These assistants can be declined in multimodal interfaces to ensure greater User eXperience (UX) in the automotive cockpit. In addition, their design can be personalized and adapted to specific uses and needs. In this paper, we report the design of an efficient virtual assistant (such as a digital twin), to increase the serenity of the journey of the occupants of a connected car, automated or not. In this study, we first aimed to investigate users' needs and preferences for vehicle assistance through brainstorming and focus group sessions. With the brainstorming sessions, 175 ideas were collected. These propositions were then evaluated in Importance and Innovation with the focus groups sessions. The results tend to show that users' concerns regarding the assistant increasingly relate to the driver's well-being, context and environment of the trip, rather than driving itself. Moreover, users want the assistant to intervene before, during and after the trip. The ideas and proposals collected will enable a better fitting of the design of our assistant prototype with user needs and preferences in terms of surface, functional, procedural and behavioral properties.

Keywords: Virtual Assistant · User Needs · Design of Automotive User Interface · Brainstorming · Accessibility

1 Introduction

In the last decade, virtual assistants have appeared progressively to support or execute user tasks in diverse domains of application. There is a rising of intelligent personal assistants which are integrated into devices like smart speakers Amazon Alexa, Google Assistant, built into a mobile operating system like Apple Siri or into a desktop operating system like Microsoft Cortana [1,2].

H. Krömker (Ed.): HCII 2024, LNCS 14732, pp. 157–170, 2024.
https://doi.org/10.1007/978-3-031-60477-5_12

These conversational agents, accessible through written commands and inputs, serve as tools for users to accomplish a wide range of tasks, including product and service research, making purchases, delivering information, and handling queries. These computer-generated applications, trained with machine learning algorithms and using natural language processing, also have the capability to acquire knowledge of user preferences as they interact over time, thereby gradually tailoring interactions to better suit the individual user [3].

Virtual assistants are present in many areas, such as in the driving context, where they can help tasks done by drivers or passengers being able to provide aids (checking navigation devices, changing the music, switching on the air conditioner, or simply talking to the driver). The role of such virtual onboard assistants may vary according to the task at hand, the level of automation, taking into account the user's cognitive state. There is no distinct demarcation of specific roles to undertake, and these assistants could fulfil a variety of roles based on the user's specific needs.

Indeed, among user's specific needs, in-vehicle assistants could have been designed to improve UX for car's occupants by doing a variety of tasks such as detecting driver distraction or fatigue and provide visual or vibrotactile alerts or engaging in conversation with the driver to keep him alert [4,5]. Likewise, if the driver is angry or sad, the agent might recommend soothing music, controlling the in-vehicle temperature, or creating a relaxing environment [6]. Furthermore, if the driver is incapable of controlling the vehicle, the onboard assistant can disengage the driver for safety by assuring a smooth transition between manual and automated driving [7] as well as communicate to other vehicles nearby (vehicle-to-vehicle) and to related authorities (vehicle-to-everything) [8].

However, agent assistance is not much thought about and even less used as a model of support for people with disabilities. Research on the subject has nevertheless been developed, such as the work of Uzan and colleagues on messaging assistance concerning the support of blind people with office interfaces [9,10]. In terms of mobility and transport, assistance technologies are designed for the general population, where they can replace tasks that drivers do or that passengers can do to help drivers. However, few projects combine these technologies for people with specific needs. However, these technical aids allow them to be independent in their daily lives and to travel in the safest possible way [11–14].

As part of interdisciplinary research, our work is about the design of an efficient virtual assistant for vehicles. Our proposal is that this digital assistant might be accessible and inclusive, a virtual figure that knows as much as possible about the driver(s) and other possible passenger(s) and about the car properties. The assistant is based on digital twins principles; i.e., the user having a digital profile stored in a cloud with information about real-world objects and is created to mimic human interaction. For this, the interaction with the users is dedicated to decision-making which, to be relevant and effective, must take into account an understanding context, the current situation but also emotional expressions, intentions, goals and tasks of the user [15].

Based on the SOLID model [16,17] it is expected that this intelligent personalized agent can "read" the profile of the user and thus knows the reasons and motivations for travel. By doing so, it "knows" when to spread useful and essential information to the user based on their needs. When dialoguing, the proactive agent, endowed with empathy by taking the other's point of view, aims to reassure, calm, or trigger interventions at the right time in order to lead users to their destination in the most serene way possible.

For the design of this onboard virtual assistant, we started to build ontologies in the form of hierarchical knowledge trees, based on use cases and on team brainstorming proposals. We quickly realized that information on the health of the driver and passenger(s), such as visual, hearing, motor or even psychological disorders, was important to be taken into account in the design of the assistant. Indeed, in order for the agent to best adapt to the user needs and constraints and offer appropriate remediations for each driving situation, the virtual assistant must know if occupants of the vehicle finds themselves in difficulty and must resort to specific solutions [18].

To define and design in detail the logic of interaction of our onboard assistant and finalize the prototype, we applied a connected object design methodology based on UX. To do this, we collected ideas, needs and recommendations based on experience of users in driving situations and environment, in manual and autonomous driving to clarify their needs and preferences for assistance in a vehicle equipped with the Virtual In-Vehicle Assistant (VIVA).

This paper presents the method used to collect ideas, recommendations and evaluation of ideas regarding users' needs and type of assistant desired in manual and automated vehicles. The results obtained are discussed and will be taken into account in the continuation of the experiments on the design of VIVA.

2 Method

The method is the making of brainstorming [19] and focus groups [20] as in current UX research with automated vehicles [21,22]. The first collective phase is a generation of proposals (brainstorming) and the second phase is the evaluation of the proposals collected during the first phase (focus groups). The participants who responded to the call for participation were unaware of the type of user task (brainstorming or focus groups). In the design process, taking into account the UX is mandatory for avoiding slips and mistakes [23,24] but also for the completion of knowledge and of Know-how about the future of things: the functions, processes and procedures about the innovative things being designed.

2.1 Brainstorming

Participants. A total of 11 participants (5 female and 6 male, mean age: 32) took part in the three sessions of brainstorming. All participants were French speakers, had driving license and a driving experience. Participants were recruited in university and via social media. They signed a consent form before the beginning of the session.

Procedure. The principle of brainstorming, which consists of bringing together a group of users so that they propose spontaneous ideas, solutions and recommendations, which are then sorted and evaluated, is to make the ideas visible to all participants to facilitate divergent thinking, that is to say the generation of associations and ideas and the sharing of ideas around the problem concerned [25].

The sessions lasted approximately two hours. The participants were invited to sit around a U-shaped table. Each brainstorming session was led by a facilitator/moderator. The role of the session leader was to present and explain the subject, displayed on a large screen with slides, as well as to describe the rules for expressing ideas (free expression, absence of judgment). For the three sessions, the questioning and exchanges between participants successively concern (i) the type of assistance, (ii) the why of the need for this assistance and finally (iii) the how of the assistance, for the future of vehicles, including autonomous vehicles.

The moderator asked participants to announce their ideas out loud and transcribe them using a pen on Post-its provided (one idea per Post-it). These annotations were systematically discussed with the group of participants and the research team, to explain their ideas, discuss and complement those of others. The experimenters facilitated the group's production of ideas (prompted if necessary) and systematically synthesized the ideas produced for each theme. Of importance, the facilitator is strictly following the method of clinical interview by avoiding eliciting the desired responses and using body language (such as leaning forward or backward, variation in gaze intensity, eye movements, facial expressions such as frowning and raising eyebrows, arm and hand positions. All of these can be inspected by participants looking for patterns indicating interest or disinterest). To do so, the facilitator must engage in finely tuned back-and-forth interactions, moving forward or backward toward the currently active participant depending on the proximity of their speech to the hidden topic of the brainstorming. All of these can be inspected by participants looking for patterns indicating interest or disinterest" [26]. A co-facilitator collected the Post-its and hung them on the wall while organizing according to the KJ method [27]. This method consists of organizing and grouping the collected ideas by theme into clusters, like a mindmap, in order to have a global vision of all ideas.

In the final phase, the real topic of brainstorming is delivered to the participants before a summary of the results that are the theme of the groups of Post-its being hung on the wall (Fig. 1).

Data Analysis. Each idea formulated on Post-its are collected, numbered and reported identically as written; but spelling errors and abbreviations are corrected. For instance, Post-it number 123 (PI-123) has the following written content in capital letters "DIGITAL RECOGNITION (BORROWED) OF THE DRIVER ON THE STEERING WHEEL => IMPOSSIBLE TO UNLOCK TEL" is an inscription which contains respectively two propositions, two spelling errors and an abbreviation. It is corrected in PI-123a - "Fingerprint recognition of the driver on the steering wheel" and PI-123b - "Unable to unlock the phone using fingerprint recognition".

Fig. 1. Ideas generated by participants during a brainstorming session, participants transcribed an idea on a Post-it which was subsequently hung on the wall.

We collected 218 ideas on post-its, that were reduced to a set 175 independent propositions overall. They were analyzed and classified into a category representing the idea it conveys. Each of the ideas was coded as:

- Assistance Related - AR (68 ideas) when the proposal explicitly concerns assistance (Example: PI-9 - "Car that prepares for arrival, to avoid a too harsh transition between the interior of the car and the exterior. (at the level of the atmosphere, sensory)").
- Off Topic - OT (28 ideas) when the proposal does not concern assistance or any of the assistance functions (Example: PI-1 - "Rather than the car, let's take the bus").
- INDetermined - IND (79 ideas) when the proposal concerns the vehicle and could implicitly concern the assistance which would justify the behavior of the vehicle or make suggestions (Example: PI-6 - "Modular configuration. Coffee maker, children's things, various objects are modular. The modules can be changed over the life of the car").

In addition, Assistance Related (AR) ideas were subdivided in 8 categories (Fig. 2):
- C1 - AVR: Assistance takes care of having a responsible car. This category brings together ideas whose concern is to reduce the environmental impact of the vehicle (Example: PI-95 - "Recommendations on driving to minimize wear and environmental impact").

- **C2 - AV**: Assistance takes care of the car. Ideas in this category relate to assistant features that take care of car maintenance tasks (Example: P1-4 - "The car could charge on its own").
- **C3 - AC**: The assistance takes control. This category concerns all ideas related to autonomous decision-making of the vehicle (Example: PI-143 - "Changing the driving mode depending on the environment (snow, turns)").
- **C4 - AA**: Assist the assistance. This category encompasses ideas centered around the means available to the user to support the assistant in carrying out their tasks (Example: PI-29 - "The assistant interface should not be voice-only").
- **C5 - ACA**: Assistance that creates the atmosphere. Ideas in this category relate to features supported by the assistant to modulate and control the interior environment of the vehicle (Example: PI-9 - "Car that prepares for arrival, to avoid a too harsh transition between the interior of the car and the exterior (at the level of the atmosphere, sensory)").
- **C6 - ACAO**: Assistance that creates the atmosphere with smells. This category is identical to the previous one with the difference that it focuses on specifically sensory modalities linked to olfaction (Example: PI-15 - "Diffuse smells, smells of coffee, pain au chocolat").
- **C7 - AAC**: Assistance adapted to the driver. This category concerns personalized assistance features according to the profile of the user (Example: PI-110 - "Intelligent assistant, efficient but non-intrusive, with relevant interventions").
- **C8 - AACT**: Assistance adapted to the driver to carry out their tasks. This last category concerns features related to user productivity and the knowledge that the assistant has of the task it must accomplish (Example: PI-10 - "Active assistant even when you are outside the car, for example, to remind you that you have to leave, traffic evaluation to adjust the departure time").

2.2 Focus Groups

Participants. A total of 7 participants (2 female and 5 male, mean age: 36) took part in the two sessions of focus groups. All participants were French speakers, had driving license and a driving experience. Participants were recruited in university and via social media. They signed a consent form before the beginning of the session.

Procedure. A focus group is "a small group of people whose response to a new product is studied to determine the response that can be expected from a larger population". The method for focus groups is quite simple because participants do not have to be creative but to provide their opinions about solutions, or to ideas about innovations.

The sessions lasted approximately two hours. Participants were invited to sit around a U-shaped table. The focus group session was led by a facilitator who first presented the criteria and announced the progress of the session by showing slides on a screen as visual support for participants' tasks. Each participant had several Post-its and a pen. An evaluation sheet of the 68 ideas (Assistant

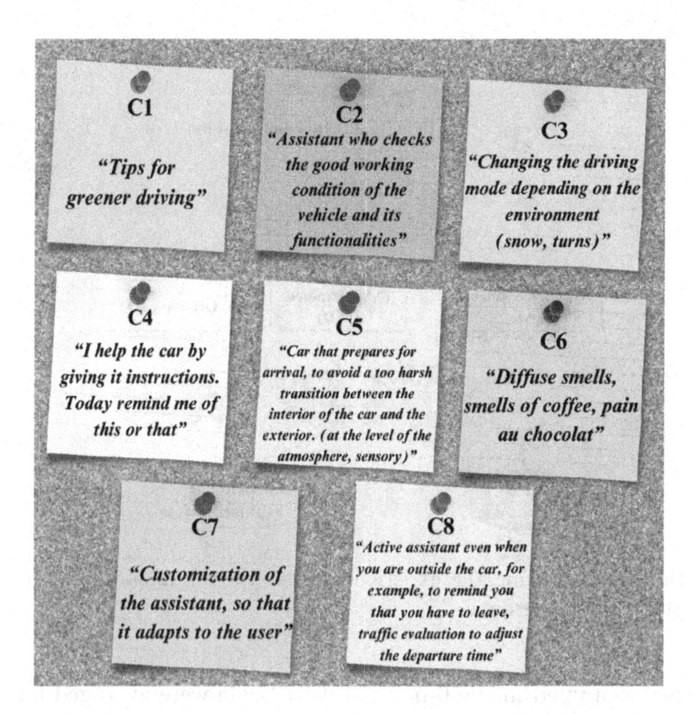

Fig. 2. Examples of collected brainstorming ideas in the 8 categories.

Related - AR) was developed from the results of the brainstorming sessions. Then, each item was presented to make the participants understand the meanings and stimulate convergent thinking.

Concerning difficulties in understanding the evaluation, the brainstorming ideas were evaluated and corrected by the focus group committee. Users rated the perceived importance and innovation of each idea on a 6-points Likert scale. In case of indecision or comprehension problems, participants could choose not to answer. The facilitator invited the participants to give their opinions, evaluations and recommendations (needs and wishes) by writing them down on the Post-its provided (one opinion or recommendation per Post-it). Participants were also asked to express comparative opinions (negative and positive). A co-facilitator collected the Post-its and hung them on the wall while organizing them by Post-it numbers.

Data Analysis. Each evaluation in Importance and Innovation and each additional suggestion were analyzed (Fig. 3).

3 Results

We specify that the criticisms and suggestions expressed in the results exclusively reflect the perceptions communicated by the participants.

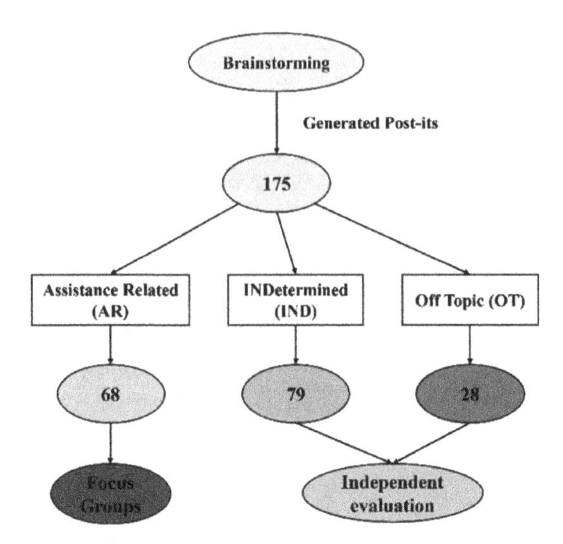

Fig. 3. Path of ideas during the experience; from their brainstorming generation to their evaluation in focus groups.

The scores obtained in the 6-points Likert Scale were averaged for each idea in Innovation, in Importance and in Importance + Innovation. The data were normalized to remove the effects of fixed effect variables (sessions and variations by participant). We performed analyzes of variance (ANOVA) to observe whether the idea category had an effect on the nature of the evaluation, in Importance on the one hand and in Innovation on the other hand.

Evaluating Innovation and Importance. Using the 6-points Likert Scale, participants were rated higher scores for Importance than for Innovation for any of the three kinds of brainstorming ideas:

- Assistance Related (AR): Mean for Importance: 3,75 vs. Mean for Innovation: 3,35
- INDetermined (IND): Mean for Importance: 4,08 vs. Mean for Innovation: 3,42
- Off Topic (OT): Mean for Importance: 3,77 vs. Mean for Innovation: 3,60

We observed that the participants were evaluated the brainstorming ideas being more Important than Innovative; $F(1, 90)=5.07$; $p=.001$. But the differences between AR, IND and OT were not significant: $F(2, 90)=.84$; $p>.43$, ns; as well as the interaction "Importance - Innovation" x "AR, IND, OT"; $F(2, 90)=1.74$; $p= .18$, ns.

In addition, there was no correlation between Importance and Innovation scores (−.274, Fig. 4), and between Importance and Innovation scores with the number of additional proposals provided by the focus group participants (.093 and .128; respectfully).

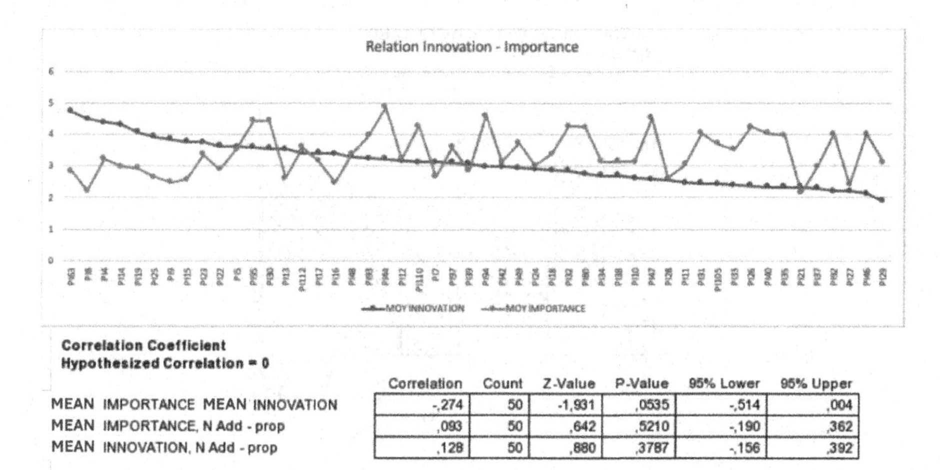

	Correlation	Count	Z-Value	P-Value	95% Lower	95% Upper
MEAN IMPORTANCE MEAN INNOVATION	-,274	50	-1,931	,0535	-,514	,004
MEAN IMPORTANCE, N Add - prop	,093	50	,642	,5210	-,190	,362
MEAN INNOVATION, N Add - prop	,128	50	,880	,3787	-,156	,392

Fig. 4. There was no correlation between Importance and Innovation scores: when the innovation score decreases, the score for Importance does not increase. Importance and innovation scores are neither correlated with the number of additional proposals provided by the focus group participants.

Evaluating Importance According to the Topic of Assistance. Evaluating the Importance of Assistance Related (AR) brainstorming ideas (Fig. 5), the ones that are of interest, there were significant differences among the 8 topic categories: $F(7, 42) = 6.15$; $p < .001$.

Importance Scores on the 6-points Likert Scale:

- C4 - AA: Assist the assistance: **4.04**
- C1 - AVR: Assistance takes care of having a responsible car: **4.01**
- C2 - AV: Assistance takes care of the car: **3.95**
- C7 - AAC: Assistance adapted to the driver: **3.43**
- C8 - AACT: Assistance adapted to the driver to carry out their tasks: **3.17**
- C3 - AC: The assistance takes control: **2.43**
- C5 - ACA: Assistance that creates the atmosphere: **2.80**
- C6 - ACAO: Assistance that creates the atmosphere with smells: **2.64**

The results presented in Fig. 5 indicate that categories C1, C2 and C4 achieve higher Importance ratings than other categories.

Evaluating Innovation According to the Topic of Assistance. Evaluating the Innovation of Assistance Related (AR) brainstorming ideas (Fig. 6), the ones that are of interest, differences among the 8 topic categories were not significant: $F(7, 42) = 5.32$; $p = .09$; ns.

Innovation Scores on the 6-points Likert Scale:

- C5 - ACA: Assistance that creates the atmosphere: **3.77**

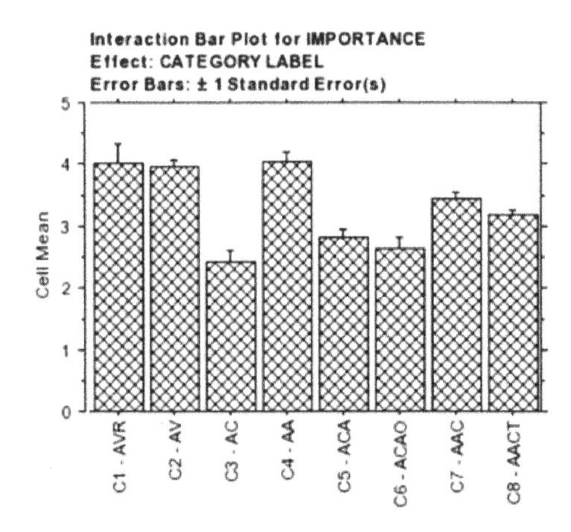

Fig. 5. Mean Importance scores assigned to ideas by category.

- C3 - AC: The assistance takes control: **3.63**
- C6 - ACAO: Assistance that creates the atmosphere with smells: **3.60**
- C1 - AVR: Assistance takes care of having a responsible car: **3.35**
- C2 - AV: Assistance takes care of the car: **3.11**
- C8 - AACT: Assistance adapted to the driver to carry out their tasks: **2.90**
- C7 - AAC: Assistance adapted to the driver: **2.84**
- C4 - AA: Assist the assistance: **2.79**

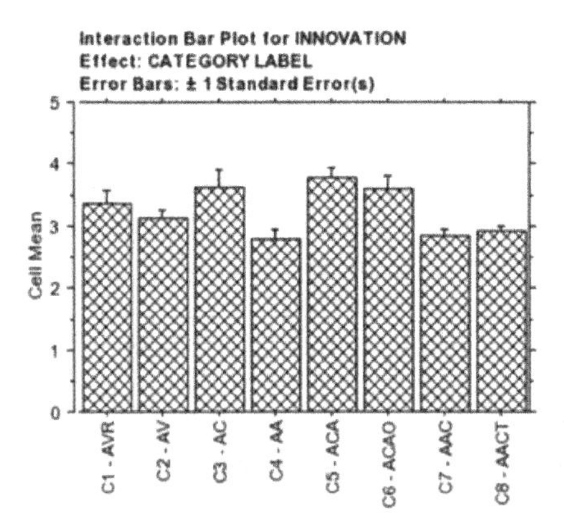

Fig. 6. Mean Innovation scores assigned to ideas by category.

The results presented in Fig. 6 indicate that categories C3, C5 and C6 achieve higher Importance ratings than other categories.

Evaluating Importance and Innovation According to the Topic of Assistance. Computing the mean scores for both Importance and Innovation (Fig. 7) show significant difference according to 8 topic categories (AVR, AV, AC, AA, ACA, ACAO, AAC and AACT): $F(7, 57) = 2.26$; $p = .04$; as well as for the interaction of the 8 topic categories with Importance and Innovation scores: $F(7, 57) = 3.53$; $p = .003$.

We observed that there is a negative correlation between Importance and Innovation according to the assistance theme (Fig. 7): being an important subject is not being important and vice versa.

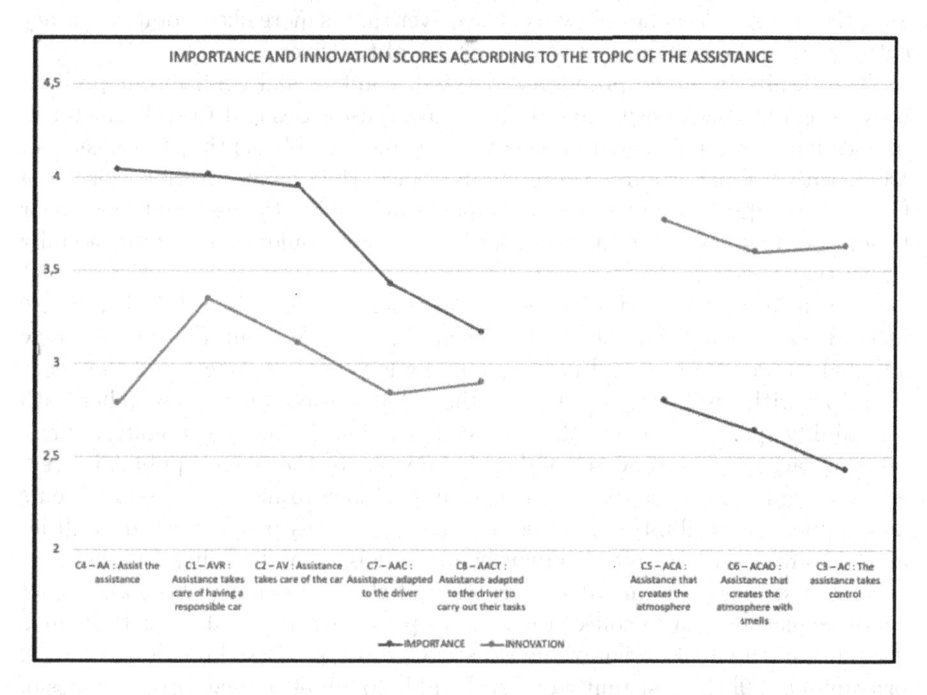

Fig. 7. As a difference with all brainstorming ideas (AR, IND, OT), the topics of Assistance Related (AR) ideas have contrasted Importance and Innovation scores: a decrease in Importance increases score in Innovation.

Evaluating How Much the User Proposals Were Already Taken into Account Within the Design of VIVA. In addition, to know which of the participants proposals were already taken into account in the design of the assistant and which of them were not, a team of four specialists will have the task to score each of the 68 proposals that were Assistance Related (AR) as already taken into account: a score of 5 if already taken into account to a score of 0 if not.

4 Conclusion and Future Directions

The findings in this study, from brainstorming and focus groups, indicate that users want an in-vehicle assistant which provides information, remediations and solutions, not on the driving systems particularly but on the context of the driving trip (ambiance in the interior, security and personalization). Driving remains a central point for the user. They want the virtual assistant to help them get to know their vehicle better so they can travel in complete safety. Vehicle occupants search for an assistant which analyzes the environment, helps them with the tasks that surround the travel and can assist them before, during and after the car trip. As a digital twin of the driver that is more like a complementary entity of the VIVA than an entity symmetrical to driver.

The results from the proposals collected, comfort and confirm our preliminary design of VIVA based on ontologies (from use cases and team brainstorming sessions). Next research step is testing and complementing the design of VIVA with UX participants testing its interface, then by an inverted Wizard of Oz technique [28] to have these participants mimicking the assistant's behavior (cognition, metacognition and pragmatics) as they would appreciate it done by the virtual assistant.

In conclusion, as part of a universal design of an onboard assistant, we focused our research on the needs of general car users, but the methodology adopted showed that several participants made proposals concerning the needs of people with disabilities. Moreover, the collaboration with a researcher with a disability quickly revealed that the construction of the needs analysis must take into account the concerns of disabled people for the general public. Indeed, when designing in-car assistants, it is of importance to meet user requirements and support accessibility for vulnerable users, such as people with disabilities and old adults with cognitive impairments. The issue of disability has therefore become a strong point in the construction of the methodology because in fact we subsequently plan to collect the ideas of people with special needs to include disability in the final design of the assistant prototype. Based on Design for All principles [9,29] the assistant will thus be able to adapt to meet diverse types of needs for any person traveling in connected and autonomous vehicles.

References

1. Pearl, C.: Designing Voice User Interfaces: Principles of Conversational Experiences. O'Reilly Media (2016)
2. Moussawi, S.: User experiences with personal intelligent agents: a sensory, physical, functional and cognitive affordances view. In: Proceedings of the 2018 ACM SIGMIS Conference on Computers and People Research, pp. 86–92 (2018). https://doi.org/10.1145/3209626.3209709

3. Cambre, J., Kulkarni, C.: One voice fits all?: social implications and research challenges of designing voices for smart devices. In: Proceedings of the ACM on Human-Computer Interaction, vol. 3, pp. 1–19 (2019). https://doi.org/10.1145/3359325
4. Large, D.R., Burnett, G., Antrobus, V., Skrypchuk, L.: Stimulating conversation: engaging drivers in natural language interactions with an autonomous digital driving assistant to counteract passive task-related fatigue. In: 5th International Conference on Driver Distraction and Inattention (DDI2017), p. 17 (2017)
5. Large, D.R., Burnett, G., Antrobus, V., Skrypchuk, L.: Driven to discussion: engaging drivers in conversation with a digital assistant as a countermeasure to passive task-related fatigue. IET Intell. Transp. Syst. **12**(6), 420–426 (2018). https://doi.org/10.1049/iet-its.2017.0201
6. BMW: Utilisez Caring Car dans votre BMW avec le système d'exploitation 7 - Tutoriel BMW (2019). https://youtu.be/nDBd9s3AphY. Accessed 16 Feb 2024
7. Mahajan, K., Large, D.R., Burnett, G., Velaga, N.R.: Exploring the benefits of conversing with a digital voice assistant during automated driving: a parametric duration model of takeover time. Transp. Res. Part F Traffic Psychol. Behav. **80**, 104–126 (2021). https://doi.org/10.1016/j.trf.2021.03.012
8. Hagiya, T., Nawa, K.: Acceptability evaluation of inter-driver interaction system via a driving agent using vehicle-to-vehicle communication. In: Proceedings of the 11th Augmented Human International Conference, pp. 1–8. (2020). https://doi.org/10.1145/3396339.3396404
9. Giraud, S., Uzan, G., Thérouanne, P.: L'accessibilité des interfaces informatiques pour les déficients visuels. In: Dinet, J., Bastien, C. (eds.) L'ergonomie des objets et environnements physiques et numériques, pp. 279–304. Hermes - Sciences Lavoisier, Paris (2011)
10. Hanine, M., Herrera, S., Chêne, D., Uzan, G.: Agent virtuel pour l'utilisation effective du web par des aveugles. In: Uzan, G., Morère, Y. (eds.) JCJC 2021 Colloque Jeunes Chercheuses Jeunes Chercheurs (2021)
11. Brinkley, J.: Implementing the ATLAS self-driving vehicle voice user interface. J. Technol. Persons Disabil. **8**, 136–143 (2019)
12. Nakagawa, Y., Park, K., Ueda, H., Ono, H.: Driving assistance with conversation robot for elderly drivers. In: Stephanidis, C., Antona, M. (eds.) UAHCI 2014. LNCS, vol. 8515, pp. 750–761. Springer, Cham (2014). https://doi.org/10.1007/978-3-319-07446-7_71
13. Pradhan, A., Mehta, K., Findlater, L.: "Accessibility came by accident" use of voice-controlled intelligent personal assistants by people with disabilities. In: Proceedings of the 2018 CHI Conference on Human Factors in Computing Systems, pp. 1–13 (2018). https://doi.org/10.1145/3173574.3174033
14. Tanaka, T., et al.: Driver agent for encouraging safe driving behavior for the elderly. In: Proceedings of the 5th International Conference on Human Agent Interaction, pp. 71–79 (2017). https://doi.org/10.1145/3125739.3125743
15. Tijus, C., Bessaa, H., Ruggieri, F., Jouen, F., Chang, C.-Y.: Digital Twins within eXtended Reality experiences (DT -XRE). Applied System Innovation (in press)
16. Uzan, G., Wagstaff, P.: Solid: a model to analyse the accessibility of transport systems for visually impaired people. In: Pissaloux, E., Velázquez, R. (eds.) Mobility of Visually Impaired People, pp. 353–373. Springer, Cham (2018). https://doi.org/10.1007/978-3-319-54446-5_12
17. Uzan, G., Pigeon, C., Wagstaff, P.: The SOLID model of accessibility and its use by the public transport operators. In: International Conference on Human-Computer Interaction, pp. 302–311. Springer, Cham (2023). https://doi.org/10.1007/978-3-031-35908-8_21

18. Lang, J., Gepner, D., Matiichak, S., Tijus, C., Jouen, F.: Digital assistive technology: the online assistance for a peaceful driving in automated and connected vehicles. In: Archambault, D., Kouroupetroglou, G. (eds.) Studies in Health Technology and Informatics. IOS Press (2023). https://doi.org/10.3233/SHTI230628
19. Osborn, A.F.: Applied Imagination: Principles and Procedures of Creative Problem Solving. Charles Scribner's Sons, New York (1957)
20. Nielsen, J.: The use and misuse of focus groups. IEEE Softw. **14**(1), 94–95 (1997). https://doi.org/10.1109/52.566434
21. Sutton, S.G., Arnold, V.: Focus group methods: using interactive and nominal groups to explore emerging technology-driven phenomena in accounting and information systems. Int. J. Acc. Inf. Syst. **14**(2), 81–88 (2013). https://doi.org/10.1016/j.accinf.2011.10.001
22. Pudāne, B., Rataj, M., Molin, E.J., Mouter, N., van Cranenburgh, S., Chorus, C.G.: How will automated vehicles shape users' daily activities? Insights from focus groups with commuters in the Netherlands. Transp. Res. Part D: Transp. Environ. **71**, 222–235 (2019). https://doi.org/10.1016/j.trd.2018.11.014
23. Norman, D.A.: Categorization of action slips. Psychol. Rev. **88**(1), 1 (1981). https://doi.org/10.1037//0033-295X.88.1.1
24. Norman, D.A.: The Psychology of Everyday Things. Basic books (1988)
25. Brown, V.R., Paulus, P.B.: Making group brainstorming more effective: recommendations from an associative memory perspective. Curr. Dir. Psychol. Sci. **11**(6), 208–212 (2002). https://doi.org/10.1111/1467-8721.00202
26. Opper, S.: Piaget's clinical method. J. Child. Math. Behav. **1**(4), 90–107 (1977)
27. Scupin, R.: The KJ method: a technique for analyzing data derived from Japanese ethnology. Hum. Organ. **56**(2), 233–237 (1997). https://doi.org/10.17730/humo.56.2.x335923511444655
28. Steinfeld, A., Jenkins, O.C., Scassellati, B.: The OZ of wizard: simulating the human for interaction research. In: Proceedings of the 4th ACM/IEEE International Conference on Human Robot Interaction, pp. 101–108 (2009). https://doi.org/10.1145/1514095.1514115
29. Lorch, R.F., Lemarié, J.: Improving communication of visual signals by text-to-speech software. In: Stephanidis, C., Antona, M. (eds.) UAHCI 2013. LNCS, vol. 8011, pp. 364–371. Springer, Heidelberg (2013). https://doi.org/10.1007/978-3-642-39194-1_43

A Study on the Effects of Different Interaction Modalities on Driving Trust in Automated Vehicles

Bo Qi, Qi Guo(✉), and Miao Liu

East China University of Science and Technology, Shanghai 200237, People's Republic of China
Guo20160820@163.com

Abstract. The growing capabilities of Intelligent vehicle interaction systems contribute to the intricacy of interaction methods, subsequently influencing drivers' trust. This study comprehensively examines prevailing interaction modes of Intelligent vehicles, employing simulated driving experiments and evaluating participants' trust levels using an automated trust scale across diverse interaction modes. The experiment introduces three specific scenarios: traffic congestion, traffic accidents, and pedestrian crossing. During the formal experiment, participants navigate a simulated urban route, encountering random instances of the specified scenarios during autonomous driving. The interaction system issues prompt when these situations arise. Four distinct interaction modes were crafted: baseline, visual, auditory, and audio-visual. The experimental findings reveal a notably heightened level of human-machine trust in the combined audio-visual interaction mode in contrast to voice and visual interactions. Furthermore, visual interaction is more effective in enhancing driver trust than auditory interaction. In visual interaction, the HUD-based visual interaction significantly elevates driver trust compared with dash-board and central control screen interfaces. This research offers insights for refining the interaction design of autonomous driving vehicles.

Keywords: Human-Machine Interaction · Intelligent Vehicles · Trust · Interaction Modes

1 Introduction

Autonomous driving technology operates vehicles through computer systems and sensors, allowing them to independently navigate and function on roads. The advancement of automated vehicles has the potential to significantly enhance people's lives and provide users with an improved driving experience [1]. As autonomous driving features become prevalent, research on the impact of driver behavior is increasingly focusing on the collaboration between humans and machines. In the current landscape of autonomous driving, effective cooperation between vehicles and drivers is essential to improve the overall driving experience and address human travel, traffic efficiency, and driving safety more efficiently. The classification of autonomous driving commonly follows the six-level system established by the Society of Automotive Engineers International (SAE International). These levels range from Level 0 (no autonomous driving) to Level 5 (fully autonomous driving) [2]. This study specifically examines vehicles at Level 2.

H. Krömker (Ed.): HCII 2024, LNCS 14732, pp. 171–181, 2024.
https://doi.org/10.1007/978-3-031-60477-5_13

Trust is a crucial factor influencing human-machine collaboration and driving safety in the context of autonomous driving [3]. The efficiency of human-machine collaboration directly impacts the safety of autonomous driving. When this collaboration efficiency is low, drivers may struggle to respond promptly to feedback from the automated system, potentially leading to driving safety issues. Insufficient trust in autonomous driving systems can result in the underestimation of their automation capabilities, keeping drivers in a state of tension during the autonomous driving process and raising doubts about the decisions made by the system [4].

Scholars have extensively researched trust in automated driving. Azevedo-Sa et al. proposed a framework to simulate drivers' dynamic trust in autonomous driving systems, estimating different trust levels. The results successfully calculated trust estimates for the continuous interaction between drivers and autonomous driving systems [5]. Clement et al. used psychological data collection to determine the level of trust humans have in autonomous driving functions. A tailored study aimed to ascertain whether advanced driving simulators could enhance consumer trust and acceptance of driving automation. They also established a connection mechanism between workload and trust [6]. Manchon et al. explored the influence of initial trust levels in highly automated driving (HAD) on driver behavior and early trust building. The results indicated that the initial level of trust in HAD affects driver behavior and further influences trust evolution [7]. Qu et al. investigated how trust changes over time and how various factors (time, trust propensity, neuroticism, and takeover warning design) collaboratively calibrate trust. The results showed that trust in automation increases in short-term interactions, decreases four months later, but remains higher than pre-experiment levels [8]. Itoh et al. developed a human-machine collaboration framework, identifying various aspects of human-machine trust based on the collaboration framework [9]. Ekman et al. studied how to create an appropriate level of user trust in advertising vehicle systems through human-machine interaction (HMI) and proposed a guiding framework for implementing trust-related factors in HMI interfaces [10]. Gao et al. proposed a dynamic trust framework that combines trust development stages and influencing factors. The framework divides trust development into four stages: predispositional trust, initial trust, real-time trust, and post-trust, analyzing key influencing factors at different stages, including operator characteristics (human), system characteristics (autonomous driving system), and situational characteristics (environment) [11].

Merritt and Ilgen posit that an operator's trust in automated systems exists as a continuum between dispositional trust and history-based trust [12]. Dispositional trust refers to the inherent level of trust individuals have in automated systems based on their own experiences, knowledge, and past encounters. History-based trust, on the other hand, develops gradually during the interaction process with the system, influenced by factors such as the system's usability, availability, and reliability. In the context of autonomous driving, a driver's trust in the autonomous driving system undergoes a dynamic transition between Dispositional and history-based trust. Initially, when drivers first encounter the autonomous driving system, they lean towards dispositional trust. As the interaction between the driver and the autonomous driving system deepens, the type of trust gradually shifts towards history-based trust. Within the historical trust states, there are three stages: initial trust, ongoing trust, and post-task trust, as illustrated in Fig. 1 [12, 13].

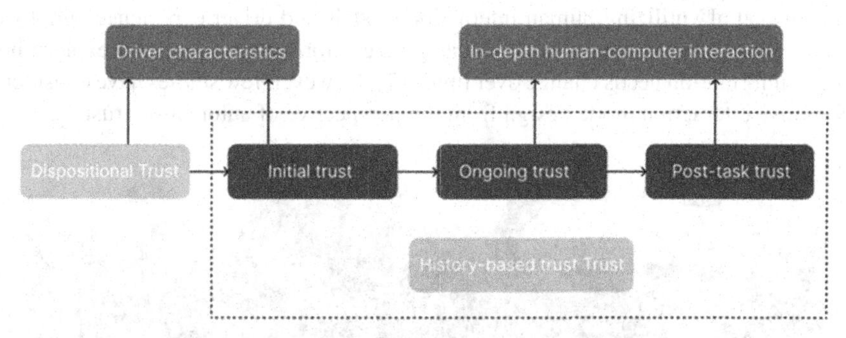

Fig. 1. The development of dynamic trust

The design of Automotive Human-Machine Interface (HMI) aims to provide an intuitive, safe, and convenient user experience, enabling drivers to effortlessly control vehicle functions, access vehicle information, and interact with the vehicle. The evolution of automotive interaction methods is a gradual process. Additionally, as automobiles are durable consumer goods, their iteration pace is slow, granting them a longer lifespan. Over an extended period of driving, the driver will form the dispositional trust and initial trust for the interaction mode.

In the long-term driving process, the driver will form the tendentious trust and initial trust for the interaction mode. As scientific and technological advancements, along with changes in human mobility, unfold, the automotive human-machine interaction (HMI) interface is gradually evolving. Based on differing interaction modes, the automotive HMI can be broadly categorized into three stages: the pure physical button interaction stage, the combination of physical buttons and screen interaction stage, and the multimodal interaction stage (see Fig. 2). Thanks to breakthroughs in Artificial Intelligence (AI) technology across various domains such as machine learning, computer vision, speech recognition, interaction, image recognition, natural language processing, and autonomous driving, the emerging field of autonomous driving has garnered widespread attention. The application of autonomous driving technology fundamentally revolutionizes the human-machine relationship within the vehicle cabin. This novel human-machine relationship demands a correspondingly innovative in-car HMI to effectively carry out various tasks in user interactions. In this technological context, multimodal interaction modes are increasingly being employed in automotive cockpit design.

In prior research, scholars have extensively investigated automotive human-machine interaction interfaces. Zhang Chi et al. proposed, based on a cognitive task analysis framework, the identification and presentation of information during takeover states to enhance safety benefits [14]. Zhou Xiaozhou et al. surveyed user preferences for HMI information features, operational modes, and envisioned interface development, constructing a user-friendly and secure HMI interface [15]. Cauffman, SJ et al. provided a comprehensive review of factors influencing the presentation of non-safety-related information on in-vehicle displays and presented a framework considering various factors determining the effectiveness of in-vehicle information delivery [16]. Some scholars have also discussed interface information content from alternative perspectives. Arun Ulahannan et al., utilizing human interaction models and driver experience, employed a longitudinal experimental design to categorize information usage and explore how drivers' information needs change over time [17]. However, few studies have considered automotive interaction mode design from the perspective of automation trust.

Fig. 2. Figure 2 From left to right: the pure physical button interaction stage, the combination of physical buttons and screen interaction stage, and the multimodal interaction stage (Image courtesy of unsplash)

This study employs a simulated driving experiment to investigate automotive interaction modes from two perspectives. The research objectives are as follows:

1. Examine which interaction mode can more effectively elevate the trust level of drivers.
2. Further explore which interaction medium can effectively elevate the trust level of drivers.

2 Methods

2.1 Participants

This study recruited 20 undergraduate and graduate students from East China University of Science and Technology (9 males, 11 females, age range from 20 to 26, Mean = 22.5, SD = 1.74). Each participant, prior to involvement, signed an informed consent form. All participants had no prior experience in similar experiments, and their vision was either normal or corrected to normal. Participants completed two rounds of experiments, and upon completion of both rounds, they received compensation for their participation.

2.2 Apparatus

This experiment was conducted in the Product Design Laboratory at East China University of Science and Technology, under normal lighting conditions (~ 300lx). The experiment utilized a 10.9-inch iPad Air 5 to simulate both the car's central control screen and dashboard. A pre-developed simulation program was installed on these devices to create a simulated driving environment. Three 23.8-inch Samsung S24R356F monitors with a resolution of 19201080 and a refresh rate of 60 Hz were employed to simulate the driving environment. The Logitech G923 TRUE FORCE kit, including a steering wheel, brake device, and gear lever, served as the simulated driving hardware. A simulated car cabin was created using the XDracing seat frame kit (580 mm–1200 mm). The entire experimental setup was placed in a dedicated enclosed room (see Fig. 3), with the simulation equipment positioned in the center of the room and the monitors against the wall to minimize external disturbances. The environmental simulation for this experiment was developed using City Car Driving.

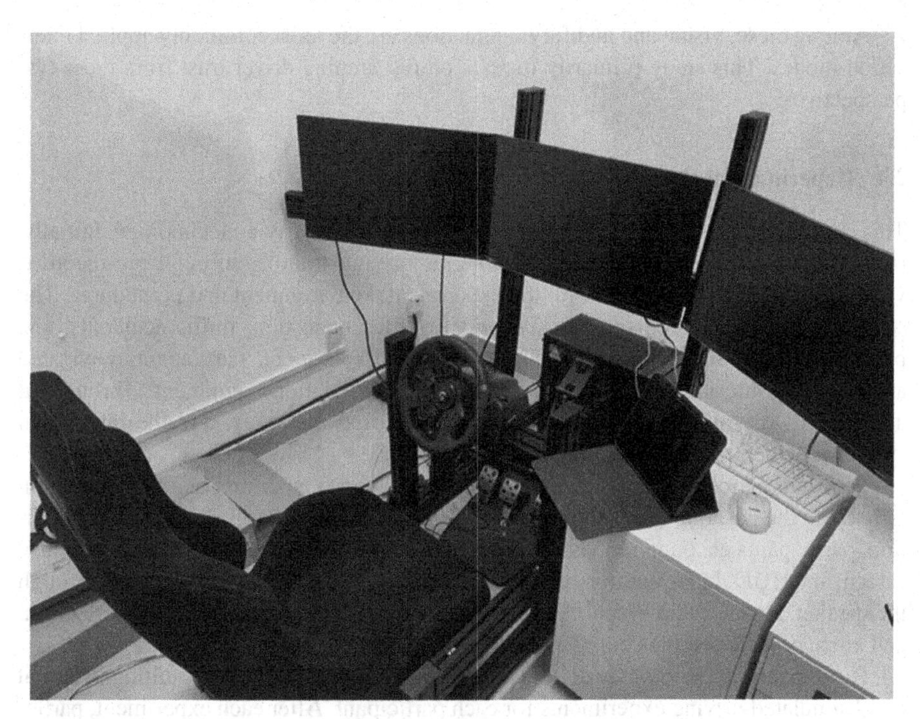

Fig. 3. Driving simulator

2.3 Interaction Model Research

This study conducts research and categorization on current interaction modes and mediums in automotive Human-Machine Interaction (HMI):

Based on different interaction mediums, interaction design can be classified into various interaction modes, representing the methods of human-system interaction. Key interaction mediums in smart vehicle interaction systems include the steering wheel, central screen, center console, doors, windows, and other components.

Depending on the interaction medium and the nature of Human-Machine Interaction (HMI), smart vehicle interaction modes can be divided into the following:

Steering wheel, doors, and windows: Physical Button Interaction (PI, Physical Stuff).

Center Console: Touch-screen Interaction (TI), Verbal Interaction (VI), System-initiative Interaction (SI) [18].

Other Interaction Mediums: Verbal Interaction (VI), System-initiative Interaction (SI), Gesture Interaction (GI).

Head-Up Display (HUD) is a rapidly advancing interaction medium. In the automotive sector, HUD typically projects information such as vehicle speed, navigation instructions, and warnings onto the car's windshield in a transparent manner. This allows drivers to access crucial information without diverting their gaze to the dashboard, contributing to increased driving safety. HUD is also a focal point of investigation in this research.

Among these, visual and auditory interactions are the most commonly applied interaction modes. This study primarily focuses on researching driver trust from these two perspectives.

2.4 Experimental Design

This experiment was designed around Level 2 autonomous driving technology. Initially, participants underwent simulated autonomous driving training through pre-recorded videos to acquaint themselves with the experimental environment and procedures. The study encompassed three specific scenarios: traffic congestion, traffic accidents, and pedestrians crossing the road. During the formal experiment, participants navigated a simulated route in urban settings using driving simulation equipment. Throughout autonomous driving, three specific scenarios occurred randomly, and the interaction system provided prompts when these situations arose.

Four interaction modes were devised: baseline, visual, auditory, and a combination of both. In the baseline condition, participants received no prompts. In the visual condition, participants encountered visual message prompts on the dashboard, central control screen, and HUD. In the auditory condition, participants received voice prompts through the speakers. In the audio-visual combination condition, participants received both visual and voice prompts on the central control screen.

Participants autonomously drove twice in each interaction mode, resulting in a total of 12 simulated driving experiments for each participant. After each experiment, participants completed a driving trust questionnaire. To gauge human-machine trust, a widely used scale developed by Jian et al. was utilized [19]. This scale, modified for the purposes of this study, exhibited good internal validity and aligns more effectively with the objectives of this research.

3 Results

Table 1 shows the mean and variance of the driving trust score of the participants after the experiment.

The intra-group Kruskal-Wallis test was conducted on the scores of the subjects under baseline conditions, and the test results showed that there was no significant difference in the dispositional trust ($H(19) = 19.470$, $p = 0.427 > 0.05$).

Table 1. Driving trust score

Group	Medium	Mean	SD
baseline	/	2.1125	0.462
Visual	Dashboard	2.7625	0.483
	Central Control Screen	2.8625	0.51
	HUD	3.9875	0.129
	Total	3.2042	0.058
Auditory	Speaker	2.45	0.121
Visual & Auditory	Central Control Screen & Speaker	4.075	0.103

3.1 Interaction Mode Trust Evaluation Results

The experimental results at the interaction mode level were analyzed (see Fig. 4). Pairwise matched-samples t-tests were conducted for trust ratings among different interaction modes. The results indicated significant differences in dispositional trust between the baseline condition and visual ($z(19) = -8.886$, $p < 0.001$), auditory ($z(19) = -2.706$, $p = 0.014 < 0.05$), and audio-visual combination ($z(19) = -14.996$, $p < 0.001$) conditions. Trust scores under the audio-visual combination interaction mode were significantly higher than those under visual interaction ($z(19) = -8.525$, $p < 0.001$) and auditory interaction ($z(19) = -10.747$, $p < 0.001$) modes. Trust scores in visual interaction were significantly higher than those in auditory interaction ($z(19) = 6.265$, $p < 0.001$).

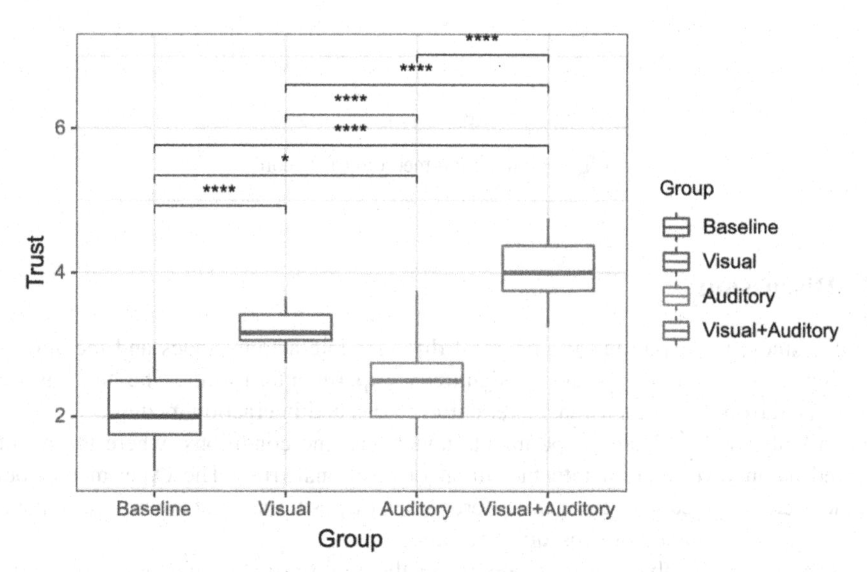

Fig. 4. Interaction mode trust score

3.2 Interaction Medium Trust Evaluation Results

Trust rating results on interaction medium were analyzed using paired-sample t-tests (see Fig. 5). The experimental findings revealed that trust scores when using HUD as a visual interaction medium were significantly higher than those with the dashboard ($z(19) = -7.374$, $p < 0.001$) and central control screen(CCS) ($z(19) = -5.674$, $p < 0.001$). However, there was no significant difference in trust scores between the dashboard and central control screen as interaction medium ($z(19) = -0.607$, $p = 0.551 > 0.05$).

Paired-sample t-tests were conducted for the three interaction medium under the baseline condition. The results showed that trust scores under all three visual interaction medium were significantly higher than those under the baseline condition ($z(19) = -4.155$, $p < 0.001$; $z(19) = -4.311$, $p < 0.001$; $z(19) = -12.677$, $p < 0.001$).

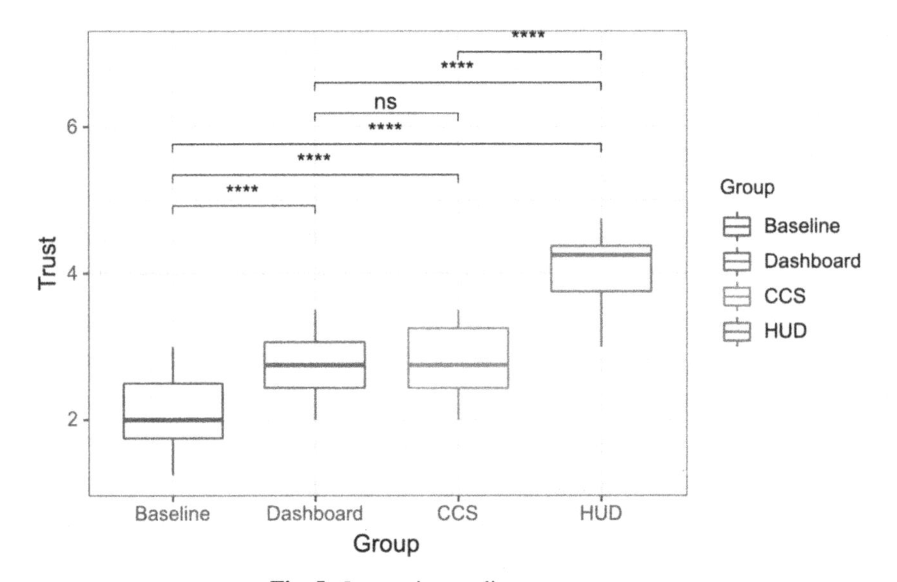

Fig. 5. Interaction medium trust score

4 Discussion

In this study, we explored the impact of different interaction modes and medium on driving trust, obtaining a series of significant experimental results. The analysis was conducted from two dimensions: interaction modes and interaction medium.

Initially, we conducted experiments under baseline conditions, where the results served as an investigation into the initial dispositional trust. The experimental outcomes demonstrated no significant differences in dispositional trust among participants, validating the accuracy of subsequent results.

In a detailed analysis, trust scores under the audio-visual combination interaction mode were significantly higher than those under individual visual and auditory interaction modes. This suggests that, in a driving environment, the combined audio-visual interaction mode might be more effective in eliciting driver trust compared to standalone visual or auditory modes. According to the multiple-resource theory [20], human information processing involves visual, auditory, cognitive, and motor response components, each with an upper limit of processing capacity. Limiting interaction to a single processing mode might lead to capacity constraints, resulting in decreased driving trust. Furthermore, visual interaction trust scores were significantly higher than auditory interaction trust scores, possibly due to the precision of information conveyed. Visual information tends to offer more intuitive and detailed insights, while auditory information might be comparatively vague or limited. In driving scenarios, drivers likely rely more on visual cues for crucial driving information, such as road conditions and traffic signals. Therefore, visual interaction may more effectively convey information, increasing driver trust in the system. This finding holds practical implications for the design of in-vehicle information interaction systems and driving assistance technologies.

Subsequently, the experiment compared the influence of different visual interaction medium, revealing that trust scores were significantly higher when using HUD as a visual interaction medium compared to the dashboard and central control screen. This indicates that HUD may have a superior effect in enhancing driver trust. The advantage lies in the fact that when drivers focus on information displayed on the dashboard and central control screen, they need to shift their gaze, diverting attention from the road conditions, whereas the use of HUD eliminates the need for such visual shifts. However, it is noteworthy that there was no significant difference in trust scores between the dashboard and central control screen. This might suggest that in certain contexts, these two interaction medium may have similar trust-building effects.

Finally, under baseline conditions, paired-sample t-tests for the three visual interaction medium and interaction modes showed that whether using HUD, the dashboard, or the central control screen, and whether employing visual, auditory, or audio-visual combination modes, all resulted in trust scores significantly higher than the baseline condition. This further emphasizes the positive role of visual interaction in enhancing driver trust and implies that choosing any visual interaction medium contributes to improving driver trust in the vehicle system.

5 Discussion

In summary, this study conducted experiments to investigate the impact of different interaction modes and medium on driving trust, yielding a series of insightful results. Regarding interaction modes, we observed that trust scores for visual interaction under baseline conditions were significantly higher than those for auditory interaction, with the audio-visual combination interaction mode showing a more pronounced effect in trust building. Furthermore, in the comparison of visual interaction medium, trust scores were significantly higher when using HUD as a visual interaction medium compared to the dashboard and central control screen. However, there was no significant difference between the dashboard and central control screen.

These findings provide valuable guidance for the future design of driver assistance systems and vehicle information interaction. Firstly, visual interaction has an advantage in enhancing driver trust, especially in scenarios involving audio-visual combination.

Secondly, HUD, as a visual interaction medium, is more likely to instill trust in drivers compared to traditional dashboards and central control screens. This holds practical significance for the design and implementation of future intelligent transportation systems and autonomous driving technologies.

Nevertheless, this study has some limitations, such as a relatively small sample size and constraints in experimental conditions. Future research could further validate and expand the study's conclusions by increasing the sample size, considering more real-world driving scenarios, and incorporating additional sensory channels. Additionally, exploring other interaction modes, such as gesture interaction and system-initiative interaction, could be pursued. Overall, this study provides important insights into understanding the impact of interaction modes and medium on driver trust, offering guidance and inspiration for the future development of intelligent transportation systems.

Acknowledgement. This study is supported by the Fundamental Research Funds for the National Natural Science Foundation of China (Grant No. 52205264) and Shanghai Pujiang Program (No. 21PJC032).

References

1. Reimer, B.: Driver assistance systems and the transition to automated vehicles: a path to increase older adult safety and mobility? Public Policy Aging Rep. **24**(1), 27–31 (2014)
2. SAE International. Automated driving levels of driving automation are defined in new SAE international standard. http://www.sae.org/misc/pdfs/automated_driving.pdf. Accessed 21 Nov 2023
3. Hancock, P.A., Nourbakhsh, I., Stewart, J.: On the future of transportation in an era of automated and autonomous vehicles. Proc. Natl. Acad. Sci. U.S.A. **116**, 7684–7691 (2019)
4. Noah, B.E., Walker, B.N.: Trust calibration through reliability displays in automated vehicles. In: Proceedings of the Companion of the 2017 ACM/IEEE International Conference on Human-Robot Interaction, pp. 361–362. ACM (2017)
5. Azevedo-Sa, H., Jayaraman, S.K., Esterwood, C.T., Yang, X.J., Robert, L.P., Tilbury, D.M.: Real-time estimation of drivers' trust in automated driving systems. Int. J. Soc. Robot. **13**, 1911–1927 (2020)
6. Clement, P., et al.: Enhancing acceptance and trust in automated driving through virtual experience on a driving simulator. Energies **15**(3), 781 (2022)
7. Manchon, J.B., Bueno, M., Navarro, J.: How the initial level of trust in automated driving impacts drivers' behaviour and early trust construction. Transp. Res. Part F Traffic Psychol, Behav, **86**, 281–295 (2022)
8. Qu, J., Zhou, R., Zhang, Y., Ma, Q.: Understanding trust calibration in automated driving: the effect of time, personality, and system warning design. Ergonomics **66**(12), 2165–2181 (2023)
9. Itoh, M., Pacaux-Lemoine, M.-P.: Trust view from the human-machine cooperation framework. In: 2018 IEEE International Conference on Systems, Man, and Cybernetics (SMC), pp. 3213–3218. Miyazaki, Japan (2018)
10. Ekman, F., Johansson, M., Sochor, J.: Creating appropriate trust in automated vehicle systems: A framework for HMI design. IEEE Trans. Hum. Mach. Syst. **48**, 95–101 (2018)
11. Gao, Z., Li, W., Liang, J., Pan, X., Xu, W., Shen, M.: Trust in automated vehicles. Adv. Psychol. Sci. **29**(12), 2172–2183 (2021)

12. Merritt, S.M., Ilgen, D.R.: Not all trust is created equal: dispositional and history-based trust in humanautomation interactions. Hum. Factors J. Hum. Factors Ergonom. Soc. **50**(2), 194–210 (2008)
13. French, B., Duenser, A., Heathcote, A.: Trust in Automation – A Literature Review. CSIRO, Australia (2018)
14. Zhang, C., Yin, G., Wu, Z.: Human-Machine Interface Research of Autonomous Vehicles Based on Cognitive Work Analysis Framework. In: Ahram, T., Karwowski, W., Vergnano, A., Leali, F., Taiar, R. (eds.) IHSI 2020. AISC, vol. 1131, pp. 99–103. Springer, Cham (2020). https://doi.org/10.1007/978-3-030-39512-4_16
15. Zhou, X., et al.: An evaluation method of visualization using visual momentum based on eye-tracking data. Int. J. Pattern Recognit. Artif. Intell. **32**, 1850016:1–1850016:25 (2018)
16. Cauffman, S.J., Lau, M., Deng, Y., Cunningham, C., Kaber, D.B., Feng, J.: Research and design considerations for presentation of non-safety related information via in-vehicle displays during automated driving. Appl. Sci. **12**(20), 10538 (2022)
17. Ulahannan, A., Jennings, P., Oliveira, L., Birrell, S.: Designing an adaptive interface: using eye tracking to classify how information usage changes over time in partially automated vehicles. IEEE Access **8**, 16865–16875 (2020)
18. Wang, R., Dong, S., Xiao, J.: Research on human-machine natural interaction of intelligent vehicle interface design. J. Mach. Des. **36**(2), 132–136 (2019)
19. Jian, J.-Y., Bisantz, A.M., Drury, C.G.: Foundations for an empirically determined scale of trust in automated systems. Int. J. Cogn. Ergon. **4**, 53–71 (2000)
20. Wickens, C.D.: Multiple resources and mental workload. Hum. Factor **50**(3), 449–455 (2008)

Investigating the Impact Factors for Trust Analysis of Autonomous Vehicle

Tianxiong Wang[1,2], Mengmeng Xu[2], Long Liu[1(✉)], Jing Chen[1], and Yuanyuan Wang[2]

[1] College of Design and Innovation, Tongji University, No. 281, Fuxin Road, Shanghai 200092, China
liulong@tongji.edu.cn

[2] School of Art, Anhui University, No. 111, Jiulong Road, Hefei 230601, China

Abstract. With the rapid development of new-generation information technology such as artificial intelligence, mobile Internet and big data, a new-generation intelligent transport system featuring automated driving will become a breakthrough in solving traffic problems. In the human-machine co-driving stage, people's trust in the automated driving system is a key element that affects the efficiency of human-machine cooperation and driving safety in automated driving, and it is crucial for drivers to maintain the high level of trust in the automated driving vehicle for driving safety. In order to explore the influencing factors of mechanism of whether people trust self-driving vehicles, the relationship between human trust and self-driving systems is analysed based on trust composition and influencing factors. Firstly, the degree of influence on trust in self-driving cars was explored in terms of attitude, perceived usability, perceived ease of use, social influence, perceived intelligence, and behavioral intention; Secondly, the structural equation model was fit-tested using validated factor analysis with structural, convergent, and discriminant validity, and the standardized coefficients of the model fit and the influence degree of the individual variables were calculated; Lastly, the model results were analysed based on the path coefficients to derive the factors influencing human trust in autonomous driving systems. The results of this study found that perceived usability, social influence and perceived ease of use have a significant positive effect on trust in autonomous driving, which is 0.864, 0.807 and 0.613 respectively. Meanwhile, the social influence could not only affect trust in autonomous driving, but also influence behavioral intention.

Keywords: Automated driving · Structural equation modelling · Trust · Impact factors · Human-computer interaction

1 Introduction

With the development of 5G, AI and other new generation information technologies, the maturity of automated driving technology is activating and reshaping the automotive industry [1, 2]. The Automated Driving (AD) is the typical representative in the new round of technological revolution, which will play an important role in changing the way of human travelling, solving road congestion and safety problems. The Google has been

H. Krömker (Ed.): HCII 2024, LNCS 14732, pp. 182–197, 2024.
https://doi.org/10.1007/978-3-031-60477-5_14

promoting driverless development since 2009 [3]. In 2015, Mercedes-Benz driverless trucks were licensed and started road testing [4]. In the same year, the Baidu innovative vehicles achieved fully automated driving in urban, ring road, and highway road conditions [5]. With the continuous development of the Internet of Vehicles and artificial intelligence technology, driverless cars have gradually become a research hotspot in the automotive industry, and the intelligent cockpit is undergoing revolutionary changes as a symbol of the intelligence of driverless cars.

When the driver is required to become the monitor of the system or stay out of the driving task, whether he can trust and delegate the driving task to autonomous driving becomes a key issue. Trust is also the driver's attitude that the autonomous driving system can help him achieve the driving task in uncertain or vulnerable situations [6]. Trust plays an important role in human-machine co-driving (trust in automation) in an autonomous driving environment, which is a key factor affecting human-machine collaboration efficiency and driving safety in autonomous driving [7]. If drivers does not trust the automatic system, they may ignore the auxiliary functions provided by the system, so as to be unable to effectively reduce traffic risks such as fatigue driving and distraction; if the drivers excessively trusts the system, they may give up monitoring the vehicle, which could lead to huge safety hazards [8, 9]. Therefore, the trust issue in autonomous driving has become a research hotspot in recent years, and the user trust level has become a necessary condition to ensure the optimal and reasonable distribution of respective responsibilities in the cooperative driving stage between humans and vehicles, especially in the adoption of new technologies [10]. Furthermore, this finding is supported by the study of Bazilsky et al. [11], who mentioned partial distrust of autonomous vehicles by those who prefer manual or partially automated driving to fully automated driving. Hence, this study shows that a high level of trust is a barrier that needs to be overcome for users to accept self-driving cars.

Most researchers who study the factors affecting trust in autonomous driving, Müller used [12] the Partial Least Squares Structural Equation Modeling (PLS-SEM) method to verify the Technology Acceptance Model (TAM) of AVs. By surveying three market groups in Europe, North America, and China, they found the compound effects of personal experience and technical norms on technology acceptance. Yuen et al. [13] studied the factors that affect users' behavioral intention to use AVs based on the integrated model of Innovation Diffusion Theory (IDT) and TAM. Although many studies have elaborated on the importance of the concept of trust between humans and machines, the existing trust models have either focused on the static trust structure of general automated systems and its related influences, or only concentrated on the influence of driver characteristics on dispositional trust and initial trust, and mainly focused on the influence of age and personality, but these studies are far from being sufficient for a comprehensive understanding of the subject and driver under human-machine co-piloting, and have not yet carried out a systematic research on them, nor have they explored in depth the influence of the design elements of automated driving on the tendency to trust.

Therefore, the purpose of this study is to explore how to better predict the adoption of self-driving cars by users, and to investigate what factors motivate people to trust self-driving cars, assist users to naturally transition from the manual driving era to the self-driving era, and improve the level of user trust, so as to give full play to its superiority in terms of efficiency, safety, and experience. To this end, the key question of this study is to explore the relationships between attitudes, perceived usability, social factors, perceived ease of use, smart perceptions and users' willingness to adopt AVs, as well as investigating what is key factors that motivate people to trust self-driving cars. Moreover, we chose the Unified Theory of Acceptance and Use of Technology (UTAUT) to explore the main factors affecting trust of AVs, which could contain more factors than the traditional TAM. The traditional TAM could only explain technology adoption behavior in terms of perceived ease of use and perceived usefulness, but the UTAUT has other social factors, So the analysis dimension is more comprehensive.

2 Review

2.1 Automated Vehicle Technology Acceptance Model

The AVs are defined in different levels according to the degree of automation [14]: level 0 is "driving automation", where the driver should be fully responsible for driving. Level 1 is called "driver assistance", where only an automated machine system is present. Level 2 is called "partial driving automation", where steering and braking as well as acceleration support are automated. Level 3 is called "conditional driving automation", in which human driving should take precedence over automated systems in many cases. Level 4 is called "high driver automation" and does not require a human driver to override in most situations. Level 5 is called "full driving automation" and does not require a human driver at all. In this study, we focus on level 3, which is conditional driving automation.

The technology acceptance model is a relatively mature theoretical model in the field of information system applications, which is proposed by Fred D. Davis of the University of Michigan in 1989 [15] and is widely used to study users' use and acceptance of new technologies. The technology acceptance model contains four core elements: perceived usefulness, perceived ease of use, usage attitude and behavioral intention. The UTAUT is an extended model of the TAM [16] and has been constructed by integrating other elements, which could include demographic factors with individual and environmental variables. However, few studies have been conducted on it that delve into the product design itself and user profiling. Therefore, this study takes an autonomous driving design perspective to examine the ways in which user acceptance of autonomous driving technologies and analyzes the impact of design elements on trust. Therefore, the model simplifies and improves the existing UTAUT, focusing the research attention on the design itself and user research, effectively avoiding the influence of external factors. To this end, the theoretical framework adopted in this study includes UTAUT and perceived value theory, and add the anthropomorphism, intelligent perception and trust factors. In fact, the anthropomorphism is an important concept in psychology [17] and it is also a good design strategy. Many artificial intelligence products have also been developed to anthropomorphize through the development of psychological characteristics to reduce consumers' perceived risks [2]. Anthropomorphism has a positive impact on technological cognition and can effectively enhance trust. When self-driving cars exhibit human characteristics, users may be more inclined to trust the functions of self-driving cars [18]. Therefore, this study incorporates it as a key indicator of intelligence perception to discover the potential relationship between it and customer trust. In addition, S Thill et al. [19] indicates that if users have a higher perception of the intelligence of an automated driving system, they tend to believe that the automated driver is able to make informed decisions in complex environments and have a higher level of trust in the utilities that the automated driver can provide, and thus perceived intelligence is also a key design factor for AVs.

Tian and Wang [14] points out that the social influence is also one of the influencing factors of trust in autonomous driving. If important people in the user's social relationship have positive opinions and are willing to use it, the user will be affected by the important people opinions and tend to perceive the AVs higher value. Therefore, this study uses the autonomous driving technology acceptance model as the theoretical basis to analyze the factors related to the improvement of trust, aiming to provide a reference for the improvement of autonomous driving technology. Furthermore, the variables is shown in Table 1.

Table 1. The measurement scale for psychological variables of autonomous vehicle users.

Variables	Test question item description	Indicators	source
Attitude	I would be happy if self-driving cars appeared	AT1	Taylor and Todd [20]
	I have a positive attitude toward self-driving cars	AT2	
	I am more in favor of the use of self-driving cars	AT3	
Perceived usefulness	Self-driving cars can make my travel more convenient	PU1	Wang et al. [21]
	Self-driving cars can improve my travel efficiency	PU2	
	Self-driving cars can improve my driving experience	PU3	
Perceived ease of use	The human-machine interface of self-driving cars is simple and easy to understand, and clearly reflects the running status of the car	PEOU1	Debernard, S. [22]
	The human-machine interface of autonomous vehicles is presented through an attractive information carrier	PEOU2	
	The information carrier of the autonomous vehicle HMI interface is reasonably laid out	PEOU3	

(continued)

Table 1. (*continued*)

Variables	Test question item description	Indicators	source
Perceived intelligence	Enhance the interaction between people and information based on touch, voice, and expressions, providing users with a more three-dimensional interactive experience	PI1	Murali [23]
	Self-driving cars have anthropomorphic expressions and sounds while driving	PI2	
	Provides intelligent adjustment mode that automatically identifies driver fatigue status	PI3	
	Intelligent adjustments to seats and steering wheels based on the user's physical characteristics	PI4	
Behavioral intentions	In my future travels, I plan to use self-driving cars	BI1	Rahman et al.[24]
	In my daily life, I would like to try a self-driving car	BI2	
	I'm interested in self-driving cars	BI3	
Social influence	Traveling in self-driving cars will make people feel distinguished and tasteful, which will bring psychological satisfaction	SI1	Madigan et al. [25]
	Important people around me think I should use autonomous driving	SI2	
	People who can influence my behavior think I should use autonomous driving	SI3	
	Someone I value wants me to use autopilot	SI4	
Trust factor	I am confident that autonomous driving can reach its destination accurately	CF1	Choi and Ji [26, 27]
	Autonomous driving improves travel safety	CF2	
	Autonomous driving is reliable when traveling	CF3	
	Autonomous driving is dependable when traveling	CF4	

2.2 The Structural Equation Model

The Structural Equation Modeling (SEM) is a statistical method to analyze the relationship between variables based on their covariance matrices [28], which could help researchers to understand and explain complex data and phenomena in a deeper way by dealing with multiple aspects such as multivariate relationships, measurement errors, causality, etc., and generally constructs the current model by describing it through the associated system of linear equations so as to result in a clearer structure. In fact, the SEM consists of two specific models [29, 30]: the first is the measurement model, which is used to establish relationships between observed and unobservable variables, and the second is the structural model, which explores path relationships between latent (unobservable) variables through path analysis. In addition, the SEM allows researchers to simultaneously consider complex relationships between multiple variables, including measurement error and causality relationships.

Thus, the SEM technology has the following advantages: (1) It can effectively measure the relationship between independent variables and dependent variables; (2) It allows to use the multiple exogenous indicator variables to represent latent variables (similar to factor analysis) and evaluate confidence and efficiency of the target variable; (3) It provides a more flexible test (i.e. measurement model) than the traditional algorithm that an indicator variable can depends on two or more underlying factors simultaneously.

3 Method

3.1 Questionnaire Design

Based on the preliminary modeling framework and research hypotheses, a survey scale was designed, which could include respondents' background information and exogenous observational variable information of latent variables. The observed variable information could include 7 latent variables of attitude, perceived usability, perceived ease of use, perceived intelligence, behavioral intention, social influence, and trust factors, and 24 observed variables are shown in Table 2. The variables were measured using the 7-point Likert scale, with 1–7 indicating from "very dissatisfied" to "very satisfied".

3.2 Questionnaire Distribution and Statistics

A questionnaire was used to conduct the study. In view of the questionnaire was about automated driving, so all participants had a driver's license and have at least three years of driving experience. If the respondents were unclear about the questions, they could ask questions and have the investigator explain them to them, and they could also communicate with the investigator during the answering process to elaborate their attitudes and opinions. Furthermore, a total of 455 questionnaires were distributed, and a total of 420 questionnaires were recovered, and removing 29 invalid questionnaires and finally obtaining 391 valid samples, with an effective recovery rate of 93.10%. Specifically, the basic demographics are shown in Table 2.

Table 2. The profile of demographic characteristics of respondents.

Project	Options	Number	Percentage	Standard Deviation
Genders	Male	180	46.04	0.835
	Female	211	53.96	
Age	20–30	206	52.69	0.711
	31–40	120	30.69	
	41–50	65	16.62	
Educational level	High School	57	14.58	0.756
	Undergraduate	203	51.92	
	Graduate Student	131	33.50	

3.3 The Questionnaire Reliability and Validity Analysis

In order to verify the validity of the results, the SPSS software was applied to test the reliability and validity of the questionnaire. The reliability mainly refers to the internal consistency between the items, and also mainly refers to the stability of the results. The valid refers to the validity of the questionnaire, which refers to the degree and accuracy to which the items can correctly reflect the content of the test. After collecting the questionnaires, we carried out a reliability test, and the Cronbach's a coefficient value of the 24 observed variables was 0.906, and the values of the 7 latent variables were all greater than 0.8 [31, 32], which indicates that the questionnaire has a good reliability. Furthermore, the KMO value of the 24 observed variables was 0.715, which is significantly greater than 0.6. When the KMO test coefficient is between the numbers 0 and 1, it indicates that the questionnaire has validity, and if the value of KMO is close to 1, it could indicate that the questionnaire has good structural validity, the detailed results are shown in Table 3.

Table 3. KMO and Bartlett's test.

KMO Measure of Sampling Adequacy		0.715
Bartlett's test of Sphericity	Approx. Chi-Square	916.218
	Df	276
	Sig	0.000

3.4 Validated Factor Analysis

The validation factor analysis could be used to confirm whether the model conforms to the previously proposed theory, and mainly verifies the convergent validity and discriminant validity of the model. The standardized factor loadings, combined reliability and Average

Variance Extracted (AVE) of each variable were obtained by calculating, and the results of the validation factor analysis are shown in Table 4. In the Table 4, the factor loadings of the question items corresponding to the seven latent variables are all greater than 0.8, which could prove that the question items can represent the latent variables to a large extent. Meanwhile, the AVE of the sums of the latent variables are all greater than 0.7, and the CR values of the combined reliability are all greater than 0.8, which proves that the convergent validity is more satisfactory.

Table 4. KMO and Bartlett's test.

Latent variables	Test variables	Standardized factor loading coefficients	AVE	Cronbach's a coefficients	Composite reliability (CR)
Use attitude	AT1	0.970	0.839	0.842	0.940
	AT2	0.917			
	AT3	0.858			
Perceived usefulness	PU1	0.954	0.902	0.833	0.965
	PU2	0.964			
	PU3	0.931			
Perceived ease of use	PEOU1	0.948	0.813	0.871	0.929
	PEOU2	0.947			
	PEOU3	0.804			
Perceptive intelligence	PI1	0.888	0.861	0.902	0.961
	PI2	0.922			
	PI3	0.928			
	PI4	0.970			
Behavioral intentions	BI1	0.938	0.863	0.945	0.950
	BI2	0.946			
	BI3	0.903			
Social influence	SI1	0.970	0.900	0.916	0.973
	SI2	0.887			
	SI3	0.967			
	SI4	0.967			
Trust factor	CF1	0.811	0.728	0.897	0.851
	CF2	0.914			
	CF3	0.804			
	CF4	0.845			

On this basis, in order to verify the reasonableness of the discriminant validity of the questionnaire, the correlation coefficient and the square root of AVE were used to assess the discriminant validity. When the square root of AVE of the latent factors is greater than the value of the correlation coefficient for other latent factors, it could prove that the discriminant validity is more appropriate. In the Table 5, there is a significant correlation between the seven potential variables, and the absolute value of the correlation coefficient is less than 0.65, and all of them are smaller than the corresponding square root of the AVE, which could prove that the potential variables have a certain degree of correlation, and have a more obvious degree of differentiation, so as to indicate that the discriminant validity of the data of the scale is reasonable and effective. Therefore, the overall reliability and validity of the questionnaire could meet the requirements through the analysis.

Table 5. The matrix of correlation coefficients between variables and square root of AVE

No	Factor 1	Factor 2	Factor 3	Factor 4	Factor 5	Factor 6	Factor 7
Factor 1	0.916						
Factor 2	0.406	0.950					
Factor 3	0.077	0.232	0.902				
Factor 4	– 0.051	– 0.191	0.443	0.928			
Factor 5	0.393	0.261	0.155	0.000	0.929		
Factor 6	0.289	0.103	0.406	0.178	0.403	0.948	
Factor 7	0.367	0.155	0.336	– 0.118	0.512	0.608	0.858

3.5 Constructing of Structural Equations Model

The path analysis of structural model is mainly to explore the causal and effect relationships among the potential variables. The structural equation model specifically contains 2 types of models: the first is the measurement model, which is used to establish the relationship between observed and unobservable variables; the second is the structural model, which could explore the path relationship between latent variables (unobservable variables) through path analysis. Specifically, the structural equation modelling usually consists of 3 matrix equations, namely:

$$\beta = \mathbf{A}\beta + \mathbf{T}\lambda + \boldsymbol{\xi} \tag{1}$$

$$Y = \Delta\mathbf{y}\beta + \varepsilon \tag{2}$$

$$X = \Delta\mathbf{x}\lambda + v \tag{3}$$

Equation (1) is a structural model, where the β is the endogenous latent variable; and the λ is the exogenous latent variable; A, T are the coefficient matrices; and ξ is the error vectors for each variable in the model. Equation (2) and (3) are measurement models, where Y is the observed variable of the endogenous latent variable; Δy is the matrix of correlation coefficients between the endogenous latent variable and the observed variable; X is the observed variable of the exogenous latent variable; Δx is the matrix of correlation coefficients between the exogenous variable and its observed variable; and ε, v are the measurement errors.

After model construction, model identification and model estimation are required. Model identification and model estimation provide the basis for the feasibility analysis of model building. In the process of model identification and estimation, the parameters of the model need to be estimated to test whether the existing designed model is realistic and to detect the degree of model fit. In this study, the model was calculated using SPSS software to obtain the values of the standardized coefficients of the structural model. The standardized path coefficients reflect the magnitude of correlation between the variables and the *p*-value is an indicator of significance. According to the Table 6, the column "Standardized model regression coefficients" could show the relationship of all measurements, in which the two elements of perceived availability and social influence have an impact on trust above 0.8, with systematic coefficients of 0.864 and 0.807, respectively. Thus, this result indicates that perceived availability and social influence have a high impact on the trust factor.

Table 6. The path analysis and model regression coefficients.

X	→	Y	Un – Standardized Regression Coefficients	SE	z (CR Value)	p	Standardized Regression Coefficients
Attitude	→	Trust level	0.342	0.035	4.091	0.000	0.346
Perceived usability	→	Attitude	0.383	0.054	7.045	0.000	0.426
Perceived usability	→	Perceptive intelligence	0.324	0.377	3.513	0.000	0.259
Perceived usability	→	Trust level	0.724	0.126	8.917	0.009	0.864
Perceived ease of use	→	Trust level	0.626	0.051	5.495	0.013	0.613
Perceptive intelligence	→	Trust level	0.218	0.047	4.691	0.000	0.462
Social influence	→	Behavioral intention	0.059	0.123	0.477	0.634	0.057
Social influence	→	Trust level	0.365	0.06	6.08	0.000	0.807
Trust level	→	Behavioral intention	0.944	0.307	3.073	0.002	0.415

In order to ensure the accuracy of the results, the fit of the model needs to be assessed. The fit of the model refers to an important metric to test the predictive ability of the model. Several key indicators of model fit were calculated by the SPSSAU software, and Table 7 shows the results of the key indicators of model fit. In practice, it is generally more difficult to make all metrics results fit the criteria. Therefore, the ideal state is that some key metrics are recognized as being within the desired range. Based on Table 7, the value of CMIN/df is 2.671, which is significantly less than 3, thus indicating a good fit of the model. Meanwhile, the relative fit indicators NFI, GFI, CFI and AGFI are all over 0.9, the absolute fit indicator RMSEA is 0.023, which is significantly less than 0.1, and all the other different parameters are within the standardized ranges, thus indicating that the information of the established model is reasonable and valid and the overall fit is good. Therefore, based on the analysis results of the established standardized structural equation model paths, the AV information structure model was established by calculating the path coefficients, and the results are shown in Fig. 2.

Table 7. The model-fitting indicators.

Commonly-Used Indicators	$\chi 2$	df	p	Cardinality Ratio of Freedom, $\chi 2/df$	GFI	RMSEA	RMR	CFI	NFI	NNFI
Judgment criteria	-	-	> 0.05	<3	> 0.9	< 0.10	<0.05	>0.9	>0.9	>0.9
Value	3521.014	240	0.000	2.671	0.905	0.023	0.01	0.948	0.933	0.996
Other indicators	TLI	AGFI	IFI	PGFI	PNFI	PCFI	SRMR	RMSEA 90% CI		
Criteria for judgement	>0.9	>0.9	>0.9	>0.5	>0.5	>0.5	<0.1	-		
Value	0.996	0.981	0.901	0.504	0.551	0.564	0.05	0.045 ~ 0.061		
Default Model: $\chi 2(276) = 9606.608$, p = 1.000										

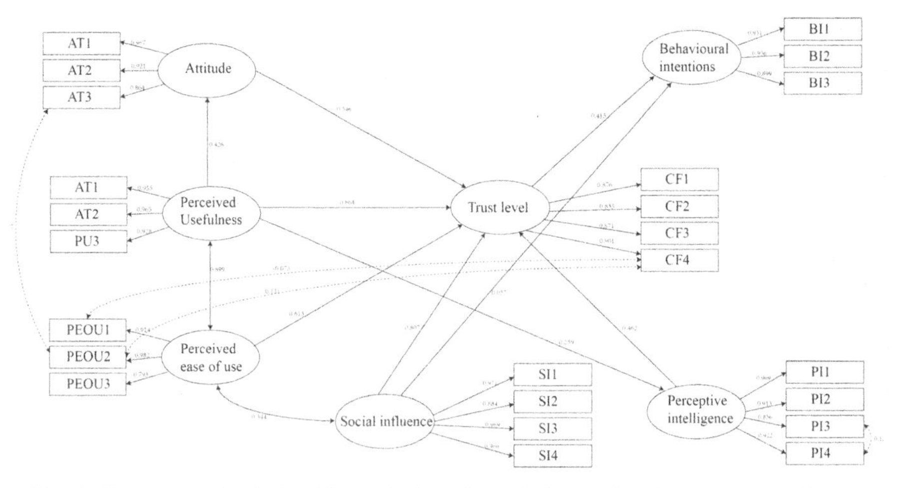

Fig. 1. The structural relationship modelling of trust influence for the autonomous driving

4 Discussions

4.1 Analysis of Factors Affecting Trust in Autonomous Driving

Based on the research ideas of the TAM, this study studies the important factors that mainly affect trust level in autonomous driving, and attempts to identify the influencing factors that affect user trust perceptions and behavioral intentions from aspects such as perceived usefulness, perceived ease of use, and use attitude. Therefore, this study can help us understand users' attitudes and expectations towards autonomous driving technology in order to improve the user experience of autonomous driving systems. According to the regression coefficient of the established structural equation model, the research results show that the performance level of perceived usefulness is 0.864, social influence is 0.807, and perceived ease of use is 0.613. Obviously, these three factors have the greatest impact on autonomous driving interaction trust and are the main influencing variables. Moreover, the comprehensive impact of attitude (0.346) and perceived intelligence (0.462) on trust in human-computer interaction for autonomous driving is relatively small, which is the secondary influencing factor. Specifically, the specific reasons are as follows:

1. The regression coefficient of perceived usefulness on trust is 0.864. This high regression coefficient shows that perceived usefulness has a significant positive impact on the willingness to trust autonomous driving. Therefore, user perceived usefulness is crucial to the adoption of autonomous driving. Perceived usefulness mainly involves the convenience, travel efficiency and driving experience of the autonomous driving system. Firstly, the convenience is an important selling point of autonomous driving technology. Users are more likely to accept self-driving technology if they feel it is easy to use and provides a more convenient transportation option. Secondly, the travel efficiency is also a key factor affecting user trust. If users believe that self-driving cars can reduce commuting time, avoid traffic jams, and thus improve travel efficiency, they will be satisfied and willing to adopt this technology. Finally, the user experience of self-driving cars will directly affect the user's driving experience, thereby affecting the user's level of trust in autonomous driving.
2. Perceived ease of use also has a significant indirect impact on users' enhanced trust in autonomous driving. If users find a self-driving car easy to use, they are likely to have a positive evaluation of the features and design of the self-driving car, thereby increasing their satisfaction with the self-driving car. Therefore, the simplicity and understandability of autonomous driving user interfaces can reduce users' cognitive load, while attractive and well-laid out user interface designs will help improve users' perceived ease of use of autonomous driving technology, thereby promoting the successful application of autonomous driving technology.
3. Users' trust in autonomous driving directly depends on social influencing factors. According to the results of the questionnaire analysis, users will be strongly influenced by the opinions and experiences of family, friends or social networks when choosing to trust the autonomous driving system. Therefore, social factors are also regarded as a key factor affecting users' trust in autonomous driving.

4. Perceived intelligence factors have a certain impact on trust building in autonomous driving technology. Intelligent perception of autonomous driving mainly refers to using some artificial intelligence, big data and human-computer interaction technologies to provide users with a more friendly experience during travel, which will also enhance users' trust in autonomous driving.
5. Attitude is also one of the factors that directly affects users' trust in autonomous driving. Differences in users' attitudes towards autonomous driving will also directly affect users' trust in autonomous driving.

4.2 Design Strategies to Enhance Trust in Autonomous Driving

Development of Autonomous Driving Functions that Fit User Experience

1. Improving users' perceived usefulness: Users' perceived usefulness is statistically significant in enhancing trust. Therefore, it is necessary to complete the research and development of functional technology from the three perspectives of convenience, efficiency and experience. Specifically, during the research and development process, we can consider starting with practical functions such as automatic summons, automatic parking, biometric unlocking, and holographic projection, paying attention to and taking care of users' travel characteristics and perception needs, and improving the practical perception of autonomous vehicles.
2. Providing more humanized interactive experience: Companies can segment the market, highlight the popularity and brand image of self-driving cars, and provide different interactive forms based on different situations to meet the personalized needs of users. In this process, more biometric identification methods are also necessary to enrich the interactive experience of autonomous driving. In addition, anthropomorphic design can make self-driving cars more approachable and attractive, increasing users' trust in self-driving cars. Anthropomorphic features can make users feel safe and reliable, allowing users to establish a more cordial and natural relationship with autonomous vehicles, making the autonomous driving system a living, emotional, thinking, considerate and reliable "partner", creating a safe, reliable and enjoyable experience, in order to better gain the understanding and trust of users.

Social Influence and Education Training

1. Utilizing social influence: This study found that people's trust in autonomous driving depends on the opinions of their surrounding relatives, friends, and neighbors about autonomous driving. Therefore, early self-driving users are encouraged to actively share their experiences and opinions on self-driving, and strengthen social influence through channels such as social media, user comments, and word-of-mouth communication, so that other potential users are more likely to trust this technology.
2. Increasing user education: Providing user training and education to help users better understand autonomous driving technology. This mainly includes training on the working principles of the systems involved in autonomous driving, intelligent sensing technology and safety, thereby improving users' trust in technology and providing users with explanations of intelligent decisions so that they can understand the motivations for autonomous driving to make certain decisions, and through scenario demonstrations and case studies, it shows the execution status of autonomous driving

technology in different scenarios, thereby helping users understand how the system's intelligent perception responds to various road and traffic situations.

5 Conclusion

In the era of AI, the relationship between humans and machines is an "elastic" human-machine relationship. In this context, human-machine integration becomes an important form of machine behavior. In this study, we analyzed the influence of psychological factors on the trust and acceptance intention of self-driving cars, and designed a human-computer interaction (HCI) trust model of AVs from six perspectives: perceived usefulness, perceived ease of use, perceived intelligence, social factors, attitude, and behavioral intention. Validated factor analysis with structural validity and discriminant validity and path analysis were used to fit the structural equation model, and key factors affecting users' trust in AVs were found.

The findings of the study are as follows: (1) Perceived usefulness, social influence and perceived ease of use have the most significant impact on trust for users with rich existing driving experience, indicating that improving users' perceived usefulness, socialization and ease of use of automated driving can largely improve the level of trust. (2) Users' intelligent perception and attitude towards automated driving can also increase the level of trust to a certain extent, therefore, how to enable automated driving to better perceive users' physiological and psychological characteristics has also become the next factor to focus on. For example, visual sensors and cameras enable the autonomous driving system to sense the driver's emotional state; in terms of entertainment, seat comfort, and ambient lighting, so that users are more willing to interact with the autonomous driving system and increase their trust in the system. (3) Socialization can influence not only trust in autonomous driving, but also user behavioral intentions. Therefore, communicating the workings and safety of an autopilot system through visual media, training, and educational materials can effectively eliminate users' misconceptions about the technology and help build trust and increase their behavioral intentions.

References

1. Wang, T., Ma, Z., Yang, L.: Creativity and sustainable design of wickerwork handicraft patterns based on artificial intelligence. Sustainability **15**, 1574 (2023)
2. Wang, T., Liu, L., Yang, L., Yue, W.: Creating the optimal design approach of facial expression for the elderly intelligent service robot. J. Adv. Mech. Des. Syst. Manuf. **17** (2023). JAMDSM0061-JAMDSM0061
3. Lutin, J.M., Kornhauser, A.L., Masce, E.L.-L.: The revolutionary development of self-driving vehicles and implications for the transportation engineering profession, Institute of Transportation Engineers. ITE J. **83,** 28 (2013)
4. Bimbraw, K.: Autonomous cars: Past, present and future a review of the developments in the last century, the present scenario and the expected future of autonomous vehicle technology. In: 2015 12th International Conference on Informatics in Control, Automation and Robotics (ICINCO), pp. 191–198. IEEE (2015)
5. Ziyan, C., Shiguo, L.: China's self-driving car legislation study. Comput. Law Secur. Rev.. Law Secur. Rev. **41**, 105555 (2021)

6. Khastgir, S., Birrell, S., Dhadyalla, G., Jennings, P.: Calibrating trust to increase the use of automated systems in a vehicle. In: Advances in Human Aspects of Transportation: Proceedings of the AHFE 2016 International Conference on Human Factors in Transportation, July 27–31, 2016, Walt Disney World®, Florida, USA, pp. 535–546. Springer (2017)
7. Rahwan, I., et al.: Machine behaviour. Nature **568**, 477–486 (2019)
8. Wintersberger, P., et al.: Second workshop on trust in the age of automated driving. In: Adjunct Proceedings of the 10th International Conference on Automotive User Interfaces and Interactive Vehicular Applications, pp. 56–64 (2018)
9. Noah, B.E., Walker, B.N.: Trust calibration through reliability displays in automated vehicles. In: Proceedings of the Companion of the 2017 ACM/IEEE International Conference on Human-Robot Interaction, pp. 361–362 (2017)
10. Gefen, D., Karahanna, E., Straub, D.W.: Trust and TAM in online shopping: An integrated model. MIS Quart., 51–90 (2003)
11. Bazilinskyy, P., Kyriakidis, M., de Winter, J.: An international crowdsourcing study into people's statements on fully automated driving. Procedia Manuf. **3**, 2534–2542 (2015)
12. Müller, J.M.: Comparing technology acceptance for autonomous vehicles, battery electric vehicles, and car sharing—a study across Europe. China, North America, Sustainability **11**, 4333 (2019)
13. Yuen, K.F., Cai, L., Qi, G., Wang, X.: Factors influencing autonomous vehicle adoption: an application of the technology acceptance model and innovation diffusion theory. Technol. Anal. Strat. Manage. **33**, 505–519 (2021)
14. Tian, Y., Wang, X.: A study on psychological determinants of users' autonomous vehicles adoption from anthropomorphism and UTAUT perspectives. Front. Psychol. **13**, 986800 (2022)
15. Davis, F.D.: Perceived usefulness, perceived ease of use, and user acceptance of information technology. MIS Quart., 319–340 (1989)
16. Venkatesh, V., Morris, M.G., Davis, G.B., Davis, F.D.: User acceptance of information technology: toward a unified view. MIS Quart, 425–478 (2003)
17. Jörling, M., Böhm, R., Paluch, S.: Mechanisms and consequences of anthropomorphizing autonomous products: the role of schema congruity and prior experience. Schmalenbach Bus. Rev. **72**, 485–510 (2020)
18. Waytz, A., Heafner, J., Epley, N.: The mind in the machine: anthropomorphism increases trust in an autonomous vehicle. J. Exp. Soc. Psychol. **52**, 113–117 (2014)
19. Thill, S., Riveiro, M., Nilsson, M.: Perceived intelligence as a factor in (semi-) autonomous vehicle UX, "Experiencing Autonomous Vehicles: Crossing the Boundaries between a Drive and a Ride" workshop in conjunction with CHI2015 (2015)
20. Taylor, S., Todd, P.: Decomposition and crossover effects in the theory of planned behavior: a study of consumer adoption intentions. Int. J. Res. Mark. **12**, 137–155 (1995)
21. Wang, S., Wang, J., Li, J., Wang, J., Liang, L.: Policy implications for promoting the adoption of electric vehicles: Do consumer's knowledge, perceived risk and financial incentive policy matter? Transp. Res. Part A: Policy Practice **117**, 58–69 (2018)
22. Debernard, S., Chauvin, C., Pokam, R., Langlois, S.: Designing human-machine interface for autonomous vehicles. IFAC-PapersOnLine **49**, 609–614 (2016)
23. Murali, P.K., Kaboli, M., Dahiya, R.: Intelligent in-vehicle interaction technologies. Adv. Intell. Syst., 4 (2021)
24. Rahman, M.M., Deb, S., Strawderman, L., Burch, R., Smith, B.: How the older population perceives self-driving vehicles. Transport. Res. F: Traffic Psychol. Behav. **65**, 242–257 (2019)
25. Madigan, R., Louw, T., Wilbrink, M., Schieben, A., Merat, N.: What influences the decision to use automated public transport? Using UTAUT to understand public acceptance of automated road transport systems. Transp. Res. Part F: Traffic Psychol. Behav. **50**, 55–64 (2017)

26. Choi, J.K., Ji, Y.G.: Investigating the importance of trust on adopting an autonomous vehicle. Int. J. Hum.-Comput. Interac. **31**, 692–702 (2015)
27. Chiou, E.K., Lee, J.D.: Trusting automation: designing for responsivity and resilience. Hum. Factors **65**, 137–165 (2021)
28. Jöreskog, K.G., Sörbom, D.: Recent developments in structural equation modeling. J. Mark. Res. **19**, 404–416 (1982)
29. Anderson, J.C., Gerbing, D.W.: Structural equation modeling in practice: a review and recommended two-step approach. Psychol. Bull. **103**, 411 (1988)
30. Kline, R.B.: Principles and practice of structural equation modeling, Guilford publications (2023)
31. Wang, T., Ma, Z., Zhang, F., Yang, L.: Research on wickerwork patterns creative design and development based on style transfer technology. Appl. Sci. **13**, 1553 (2023)
32. Wang, T., Zhou, M.: A method for product form design of integrating interactive genetic algorithm with the interval hesitation time and user satisfaction. Int. J. Ind. Ergon. **76** (2020)

Impacts of Automated Valet Parking Systems on Driver Workload and Trust

Zhenyuan Wang[1], Yizi Su[1], Qingkun Li[2,3](✉), Wenjun Wang[1], Chao Zeng[4,5], and Bo Cheng[1]

[1] The State Key Laboratory of Automotive Safety and Energy, Center for Intelligent Connected Vehicles and Transportation, School of Vehicle and Mobility, Tsinghua University, Beijing 100084, China

[2] Institute of Software Chinese Academy of Sciences, Beijing 100190, China
liqingkun@iscas.ac.cn

[3] Automotive Software Innovation Center (Chongqing), Chongqing 401331, China

[4] Henan University of Technology, Zhengzhou 450001, China

[5] Hami Vocational and Technical College, Hami 839001, China

Abstract. In the realm of automated driving, automated valet parking (AVP) systems represent a significant leap towards enhancing urban mobility and safety. While existing research has explored various aspects of AVP systems, there is a notable gap in the literature specifically addressing AVP systems in relation to their impact on driver workload and trust. This study evaluates a Level 2 AVP system implemented in a vehicle, focusing on its impact on driver workload and trust. We utilized eye-tracking, physiological monitoring, and self-reported surveys to capture driver responses during AVP operation compared to manual parking. Results indicated a trend towards reduced workload during AVP use, as suggested by eye-tracking data, and a decrease in physiological markers of stress, although these differences were not statistically significant. Driver trust in the AVP system significantly increased after hands-on experience. However, an increase in mobile device usage signaled potential issues of overreliance on automation. The findings underscore the importance of integrating human factors into AVP system design to balance workload reduction with the prevention of overreliance, ultimately enhancing driver engagement and calibrating trust.

Keywords: Automated Driving · Driver Workload · Driver Trust · Automated Valet Parking System

1 Introduction

The rise of automated driving technologies is revolutionizing urban mobility and road safety, with automated valet parking (AVP) systems emerging as a key innovation [1]. These systems, designed to autonomously park vehicles without active driver intervention, promise not only enhanced convenience for drivers but also a potential reduction in parking-related incidents, a common issue in dense urban environments [2].

H. Krömker (Ed.): HCII 2024, LNCS 14732, pp. 198–210, 2024.
https://doi.org/10.1007/978-3-031-60477-5_15

AVP's relevance is particularly pronounced in crowded urban settings, where efficient space management and reducing the risk of parking collisions are crucial [3]. By automating this task, AVP systems aim to optimize space usage and alleviate the challenges associated with manual parking, while minimizing human error and enhancing safety [4]. This aspect of AVP aligns with the broader movement towards vehicle automation, showcasing a shift in the dynamics of vehicle interaction and urban transportation, and marking a pivotal advancement in the journey towards fully automated driving.

However, the partial autonomy offered by AVP systems also introduces distinct human factors challenges [5, 6]. Despite the technological sophistication, these systems still require the human driver to remain responsible for monitoring the vehicle's performance and being prepared to take control when necessary [7, 8]. This dual demand places a significant emphasis on understanding and balancing the driver's workload and his/her trust in the AVP system.

Workload, in the context of AVP, refers to the mental effort required by the driver to monitor and potentially intervene in the automated parking process [9]. Excessive workload can lead to fatigue and decreased vigilance, increasing the risk of accidents during critical moments of transition between automated and manual control [10]. Conversely, insufficient workload may result in overreliance on the system, potentially leading to delayed response to unforeseen events.

Trust in automation, especially regarding systems like AVP, plays a crucial role in how drivers use and monitor these systems [5]. Appropriate levels of trust are essential to ensure that drivers remain engaged and ready to intervene, but without becoming overly reliant on the technology [11, 12]. Balancing this trust is complex, as it involves not only the system's reliability and performance but also the driver's perceptions, experiences, and propensity to trust automated systems [13].

Thus, the human factors challenges in AVP systems are not just about the technical operation but also about the interaction between the driver and the automation [14]. Understanding and addressing these challenges is vital for the successful implementation and acceptance of AVP systems, ensuring they enhance, rather than compromise, overall safety and efficiency in urban mobility.

While existing research has explored various aspects of automated driving, there is a notable gap in the literature specifically addressing AVP systems, particularly in relation to their impact on driver workload and trust [15]. Previous studies, mainly based on questionnaires and simulator tests, don't fully reflect the complexities of real-world AVP system interactions. This gap highlights the need for empirical research using actual vehicles equipped with AVP technology to provide more ecologically valid insights [16].

The primary aim of this study is to examine the influence of a Level 2 AVP system on driver workload and trust by adopting a Level 2 AVP system implemented in a production vehicle. In the context of increasing automation in vehicles, understanding these aspects is critical for ensuring that such systems are both effective and align with user needs. Specifically, the study seeks to evaluate how interaction with an AVP system affects the drivers' workload and their confidence in the system's capabilities.

This study's significance lies in its contribution to the understanding of human factors in AVP systems. By utilizing a real vehicle equipped with Level 2 AVP technology, it enhances the ecological validity of the research, providing insights that are directly applicable to real-world usage. The findings are expected to inform the development of user-centered AVP systems, guiding design decisions to foster appropriate trust, reduce cognitive workload, and enhance safety and user satisfaction in the emerging landscape of automated transportation.

The paper is organized into several key sections: Following this introduction, Sect. 2 details the experimental design, including the AVP system used, participant recruitment, data collection techniques, and analytical methods. Section 3 presents the findings related to cognitive workload, trust, and other relevant human factors metrics. This is followed by Sect. 4, which interprets the results, highlights their implications, and considers the study's limitations and areas for future research. The paper concludes with Sect. 5, summarizing the key contributions of the study to the field of automated driving and human factors.

2 Method

2.1 Experimental Platform

The study focuses on the AVP system integrated into the Xpeng P7, which represents a sophisticated Level 2 automation feature. This system combines advanced cameras and an Inertial Navigation System for environmental perception, along with deep learning-based software algorithms for mapping the parking area. These technical elements enable the AVP to record parking routes and autonomously replicate them with high accuracy. The fusion of multiple sensor inputs facilitates complex tasks such as obstacle avoidance and navigation in tight spaces, crucial for effective functioning in varied parking scenarios.

The relevance of the AVP system to this study lies in its operational dynamics. While it reduces the physical demands of parking, the system still necessitates continuous driver monitoring, introducing a unique cognitive aspect to the driving experience. This study aims to explore the impact of this dual nature of the AVP system – reducing operational workload while requiring vigilant supervision – on the driver's workload and trust. Understanding this balance is essential in evaluating the real-world effectiveness of AVP technologies and their influence on driver behavior and perception.

2.2 Experimental Design

This study employed a within-subjects experimental design with two independent variables: parking mode (i.e., manual driving versus AVP automated parking) and scenario difficulty (high difficulty versus low difficulty), as shown in Table 1 Tscenarios were strategically chosen to assess the AVP system's efficacy under different parking challenges, reflective of real-world conditions [17].

The high difficulty scenario involved navigating narrow spaces and avoiding additional obstacles, demanding precision and attention, while the low difficulty scenario offered ample parking space and fewer obstructions, representing a more common and less challenging parking situation [4]. Both scenarios were set in actual parking lots of a building, starting outside the lot and ending in a parking spot, thus mimicking typical parking experiences encountered by drivers [18].

Participants completed a total of six parking trials, two in each mode of parking, across both scenario difficulties. The sequence of these trials was carefully arranged to balance any ordering effects, ensuring a fair evaluation of the AVP system's impact on the driver's cognitive workload and trust. In each trial, a pedestrian crossing event was incorporated to investigate the driver's reaction to unexpected situations, adding an element of realism and unpredictability to the experiment.

Table 1. An example of the trial setting

No	Mode	Scenario Difficulty
1	Manual driving	Low difficulty
2	AVP driving	Low difficulty
3	Manual driving	Low difficulty
4	AVP driving	High difficulty
5	Manual driving	High difficulty
6	AVP driving	High difficulty

2.3 Measurements

To comprehensively assess the impact of the AVP system on driver workload and trust, a multifaceted measurement approach was employed. The selection of specific metrics was guided by their established efficacy in observing workload and trust during driving tasks.

Eye movement: Pupil diameter and blink rate were recorded using a Tobii eye tracker. These eye movement indicators are widely recognized for their sensitivity to changes in cognitive workload. Pupil dilation typically correlates with increased mental effort, while blink rate can reflect changes in stress and attention levels [19]. Thus, they provide a non-invasive yet effective means to gauge the cognitive state of drivers during the operation of the AVP system.

Face Orientation: The orientation of the driver's face, measured in terms of pitch, yaw, and roll angles, was captured using OpenCV computer vision algorithms applied to video footage. This measurement aids in understanding the driver's focus and attention allocation, especially towards in-vehicle displays, which is crucial in evaluating their engagement and monitoring behavior while using the AVP system.

Heart Rate Variability (HRV): HRV was monitored using a BIOPAC MP150 module to record ECG data. HRV metrics include time domain measures and frequency domain power spectra. Time domain measures comprise SDNN (Standard Deviation of NN intervals, reflecting overall heart rate variability), RMSSD (Root Mean Square of Successive Differences, indicating parasympathetic activity), and pNN50 (the proportion of NN50 divided by total NNs, representing rapid changes in heart rate). Frequency domain power spectra analyze the distribution of power into frequency components, such as power in low-frequency (LF) and high-frequency (HF) bands. These cardiovascular responses are directly linked to workload, making HRV a valuable metric in assessing the physiological impact of AVP system usage on drivers [20].

Self-Report Surveys: Following each trial, participants completed customized questionnaires designed to measure perceived workload (using the RAW-TLX scale), fatigue, stress, situation awareness, and trust in the AVP system. These self-reports provide subjective insights into the drivers' experiences, complementing the objective physiological and behavioral data [16].

Each of these measures offers a unique perspective on the driver's cognitive and emotional state, creating a holistic view of their interaction with the AVP system.

2.4 Participants

This study engaged five male university students in their early 20s, a group characterized by their potential familiarity with automated driving technologies. This demographic choice, while beneficial for initial exploratory research, presents limitations in terms of generalizability. The homogeneity of the participant group, primarily in terms of age, gender, and educational background, may influence their perception and interaction with the AVP system. Their likely familiarity with technological advancements, particularly in automated driving, could predispose them to higher baseline trust and different cognitive workload management than a more diverse population.

2.5 Procedure

The procedure of the experiment was designed to ensure clarity, safety, and consistency. It began with obtaining informed consent from the participants, ensuring they understood their involvement in the study. Once consent was established, participants were equipped with sensors to collect physiological data.

Participants underwent standard training, typical for driving experiments, focusing on familiarizing them with the Xpeng P7 vehicle and its AVP system. They were also briefed on the parking environment where the trials would take place.

To prioritize safety during the experiment, several measures were implemented. The vehicle's speed was kept within a safe limit, suitable for the controlled parking lot environment. Additionally, all participants were required to wear seat belts while driving. An experiment operator was always present in the vehicle, seated next to the driver. This operator closely monitored traffic conditions and was prepared to take immediate control of the vehicle if necessary, ensuring an extra layer of safety.

After the trials were completed, and all data were recorded, the procedure concluded with a post-experiment debriefing. During this session, participants had the opportunity to discuss their experiences and provide feedback on the experiment.

2.6 Data Analysis

In the pre-processing stage, eye tracker data were processed using ErgoLAB 3.0, a specialized software for analyzing eye movement data. For the ECG data, Kubios HRV Scientific software was employed. Following pre-processing, the data underwent statistical analysis using Python.

3 Results

3.1 Eye Tracking Metrics

The evaluation of eye tracking metrics revealed distinct trends when comparing AVP with manual driving, despite the lack of statistical significance in the differences observed, as shown in Fig. 1.

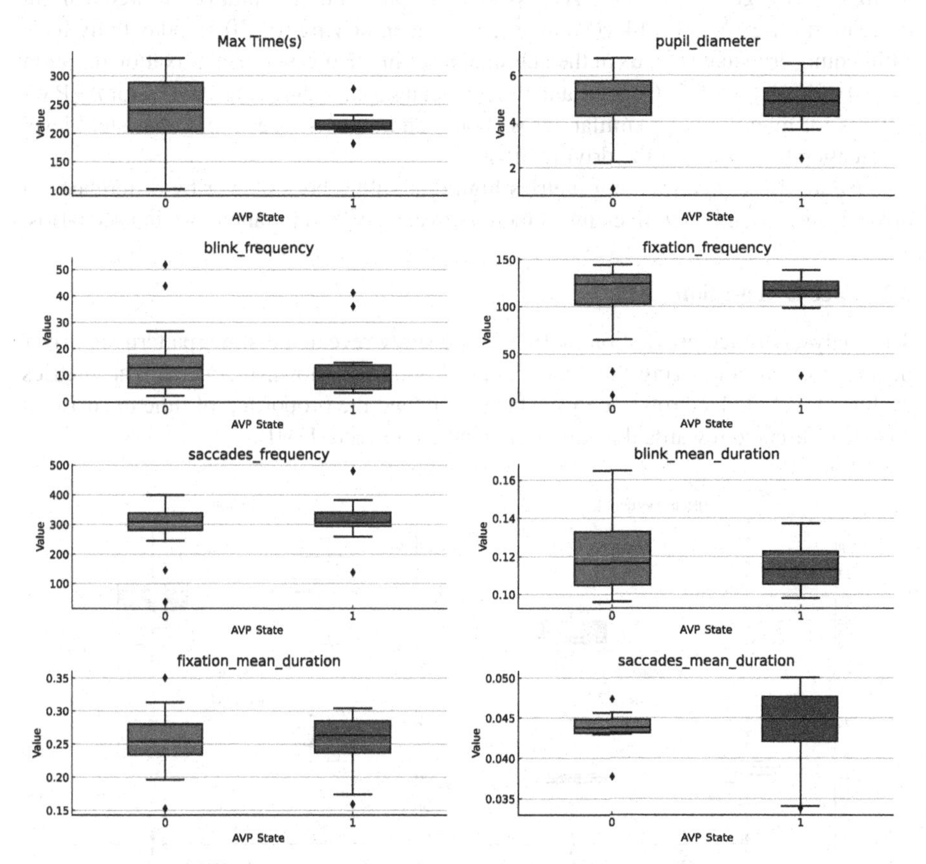

Fig. 1. The boxplots of eye tracking metrics comparison

In terms of pupil diameter, a key indicator of workload, the average measurement during AVP was 4.810 mm, slightly lower than the 4.851 mm recorded during manual driving. This trend suggests a potential decrease in cognitive workload with the use of AVP, as indicated by the smaller pupil diameter.

The blink rate exhibited a similar pattern, averaging 13.751 blinks per minute under AVP conditions compared to 15.143 blinks per minute during manual control. This reduction in blink rate aligns with the notion of reduced stress or lower workload in the context of AVP.

Regarding saccade frequency, which reflects the driver's visual scanning behavior, there was an increase to 311.830 during AVP use, from 293.087 in manual driving.

This higher frequency during AVP operation may indicate a more active scan of the environment, which is a crucial aspect of maintaining situational awareness in automated driving scenarios.

Complementing these findings, additional eye tracking metrics also presented notable differences. The fixation frequency was marginally higher in AVP (109.803) than during manual driving (109.230), possibly reflecting a consistent level of monitoring and engagement with the AVP system. The mean blink duration showed a slight decrease during AVP (0.114 s) compared to manual driving (0.120 s), potentially indicating quicker visual refocus in the automated setting. However, mean fixation duration (0.250 s for AVP vs. 0.256 s for manual) and mean saccade duration (0.043 s for AVP vs. 0.044 s for manual) were similar across both conditions, suggesting a consistent level of attention regardless of the driving mode.

Overall, these eye tracking metrics highlight subtle but noteworthy differences in driver behavior and cognitive engagement between AVP and manual driving scenarios.

3.2 Face Orientation

The analysis of face orientation metrics in the study revealed distinct patterns in driver behavior when comparing AVP to manual driving, as shown in Fig. 2. The metrics evaluated include face roll, face yaw, face pitch, and the proportion of time the driver's gaze was directed towards the Human-machine interface (HMI).

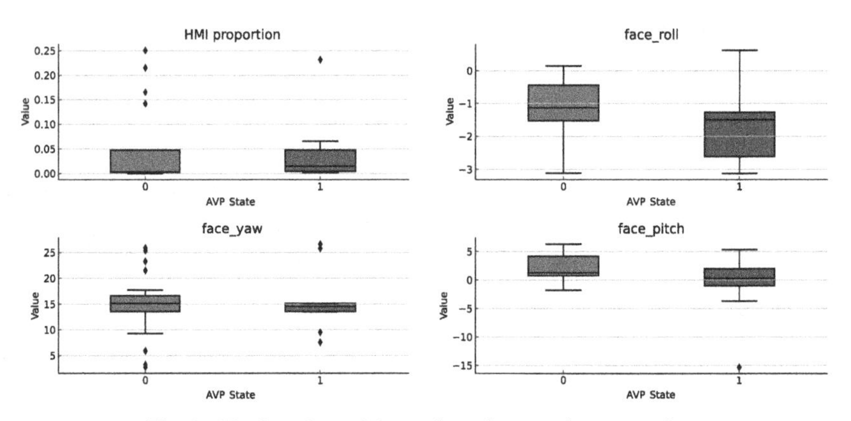

Fig. 2. The boxplots of face orientation metrics comparison

The face roll angle, reflecting the tilting motion of the head, showed a slight change from an average of -1.062 during manual driving to -1.681 in AVP. Similarly, the face yaw angle, indicative of left-right head movements, exhibited a minor increase from 14.797 in manual driving to 15.716 in AVP. However, statistical analysis via t-tests revealed that these differences in face roll ($t = 1.14, p = 0.273$) and face yaw angles ($t =$ -$0.31, p = 0.363$) were not statistically significant, suggesting consistent head movement patterns across both driving modes.

A more notable finding was observed in the face pitch angle, which measures the vertical orientation of the face. During AVP, the average face pitch angle significantly shifted to -1.308, compared to 2.381 in manual driving, with a t-test showing this difference to be significant ($t = 1.57, p = 0.015$). This downward orientation during AVP use suggests an increased focus on in-car displays, possibly reflecting the driver's engagement with the AVP system or monitoring of vehicle feedback.

Regarding the HMI proportion, which represents the proportion of time the driver's gaze was focused on the in-car displays, a slight decrease was observed during AVP (0.0426) compared to manual driving (0.0492). The t-test for this metric showed a borderline significance ($t = 0.19, p = 0.085$), indicating a trend where drivers engaged slightly less with the HMI during AVP.

In summary, while the roll and yaw angles of the head did not show significant differences between AVP and manual driving, the significant change in the pitch angle and the trend in HMI proportion highlight how drivers may interact differently with vehicle controls and displays during the use of AVP systems.

3.3 Heart Rate Variability

The HRV analysis provides significant insights into the physiological responses of drivers during AVP compared to manual driving. A summary of the HRV metrics for both driving conditions is presented in Table 2 below:

Table 2. Summary of the HRV metrics results

HRV Metric	AVP	Manual Driving	*p*-value
Mean RR (ms)	974.94	869.46	0.044
SDNN (ms)	36.5	46.08	0.016
Mean HR (bpm)	61.67	70.11	0.051
RMSSD (ms)	39.32	46.11	0.053
pNN50 (%)	41.17	81.75	0.039
LF power (ms2)	18.63	26.1	0.016
HF power (ms2)	484.15	1034.06	0.037

From the table, it is evident that several HRV indicators showed significant differences between AVP and manual driving. The Mean RR interval was longer during AVP, with a statistically significant increase ($p = 0.044$), suggesting a more relaxed state with lower heart rates. The SDNN, a measure of overall HRV, was lower in AVP ($p = 0.016$), indicating a reduction in heart rate variability and potentially a lower level of stress. Similarly, the RMSSD and pNN50 values, which are indicators of short-term variations in heart rate, were lower during AVP (*p*-values: 0.053 and 0.039, respectively).

Moreover, the power in the LF and HF bands, which are associated with sympathetic and parasympathetic activities, respectively, also showed significant differences. Both LF and HF powers were lower in AVP (*p*-values: 0.016 and 0.037), suggesting a reduced activation of both sympathetic and parasympathetic nervous systems.

These results collectively indicate a trend towards a more relaxed and less variable cardiac response during AVP use compared to manual driving. The lower mean HR, reduced variability (SDNN), and decreased power in both LF and HF bands in AVP suggest a decrease in physiological stress and cognitive workload.

3.4 Self-reported Surveys

This section delves into the self-reported trust and correlation analysis between various survey measures and the AVP system.

Trust Score Comparison. The Self-Trust Scale (STS-AD) [21] was employed to measure participants' trust in the AVP system. A comparison of trust scores before and after experiencing the AVP system showed a significant increase, with the mean trust score rising from 0.598 before use to 0.805 after use. This considerable change was supported by a high t-value of 9.938 and a p-value of less than 0.001, indicating a statistically significant enhancement in trust due to the firsthand experience with the AVP system.

Correlation Analysis. A correlation analysis was conducted to explore the relationships between trust in the AVP system and various other measures post-experience. The heatmap in Fig. 3 illustrates the correlation coefficients ranging from -1 to 1, where 1 represents a perfect positive correlation, -1 a perfect negative correlation, and 0 no correlation.

	Driving Year	AD Experience	DSI	MDSI Dissociative	MDSI Anxious	MDSI Risky	MDSI Angry	MDSI High Velocity	MDSI Distress Reduction	MDSI Patience	MDSI Careful	TIA Pre	TIA Post
STS_AD	-0.61	0.12	-0.66	-0.59	-0.12	-0.98	-0.85	-0.82	0.07	0.22	0.64	0.61	0.95
RMSE	-0.69	0.14	-0.72	-0.67	0.34	-0.45	0.21	0.19	-0.78	-0.55	-0.35	-0.33	0.25
KSS	-0.44	0.53	-0.83	-0.83	0.02	-0.55	-0.15	-0.07	-0.79	-0.55	0.04	0.15	0.50
DALE	0.69	-0.21	0.94	0.68	-0.18	0.84	0.52	0.51	0.43	0.09	-0.21	-0.27	-0.84
SART	0.23	-0.20	0.39	0.37	0.34	0.78	0.98	0.91	-0.34	-0.39	-0.85	-0.85	-0.88
DSSQ	0.70	0.50	0.46	0.01	-0.27	0.81	0.80	0.91	-0.52	-0.73	-0.40	-0.29	-0.85

Fig. 3. The heatmap of correlation analysis results.

From the heatmap, we can observe that post-experience trust (STS-AD) has a notably high positive correlation with metrics such as the measure of self-reported mental effort (RSME) and the measure of driver activity load index (DALI), with correlation coefficients of 0.69 and 0.64, respectively. This suggests that participants who reported lower mental effort and activity load also reported higher trust in the AVP system.

Conversely, trust appears to be negatively correlated with measures such as Karolinska Sleepiness Scale (KSS) measuring drowsiness and Situational Awareness Rating (SART), with correlation coefficients of -0.44 and -0.23, respectively. Interestingly, the highest positive correlation with trust was found in Driver Stress and Workload Questionnaire (DSSQ), at 0.7, indicating that drivers who felt less stress and workload when using the AVP system reported higher trust levels.

4 Discussion

4.1 Driver Workload and Trust

The results of our study provide a nuanced understanding of the cognitive workload and trust in the context of AVP systems. The eye-tracking measures, including pupil diameter and blink rate, alongside the HRV metrics, collectively pointed towards a reduction in cognitive workload when the AVP system was active. Although these differences did not reach statistical significance, possibly due to the limited sample size, the trends are indicative of the intended effect of AVP systems to alleviate the cognitive demands of parking.

A particularly striking outcome of this study was the substantial increase in trust scores following the experience with the AVP system. The statistically significant rise in trust, as quantified by the STS-AD, underscores the importance of direct interaction with technology in shaping users' perceptions and confidence. This boost in trust is essential for the adoption of AVP technology, as trust is a critical facilitator of technology use.

The face orientation data provided additional insight, suggesting that drivers were more inclined to focus downward, presumably towards in-car displays, when the AVP system was engaged. This finding could be indicative of drivers monitoring the AVP system or seeking feedback from the vehicle during the automated parking process. While it is positive to see engagement rather than disengagement during automation, it is essential to distinguish between monitoring that contributes to situational awareness and that which may indicate uncertainty or over-checking due to a lack of confidence in the system.

These results together highlight the delicate balance required in automated system design. On the one hand, automation should reduce the driver's operational burden to a level that prevents overload and stress; on the other hand, it should not foster overreliance or complacency. The enhanced trust in the AVP system post-exposure is a double-edged sword: while it is a testament to the system's usability and potential for reducing driver workload, it also raises concerns about the potential for overtrust and misuse.

Future AVP systems should be designed to maintain driver engagement and support appropriate trust calibration. This involves not only the automation's reliability and performance but also the user interface and feedback mechanisms that keep the driver informed without inducing unnecessary stress or workload. By doing so, AVP systems can ensure safety and user satisfaction, aligning with the ultimate goals of human-centered automation in transportation.

4.2 Implications for Design and Overreliance

The observed increase in mobile device usage during the operation of the AVP system brings forward critical concerns regarding potential overreliance and complacency. Such behavior is indicative of a shift in the driver's attention away from the primary task of driving, which, while less critical during the automated parking process, may become a hazardous habit if generalized to other driving contexts. This underscores the need for AVP systems to incorporate design features that manage the driver's engagement without fostering overreliance.

To mitigate these risks, AVP system designers could consider implementing intermittent engagement tasks for the driver or alerts that ensure the driver remains prepared to resume control if necessary. Additionally, clear indicators of system status and limitations could help in maintaining a realistic perception of the system's capabilities, potentially reducing the incidence of over trust.

Furthermore, our findings contribute to the broader discourse on the integration of human factors into the design of automation technologies. It is imperative that these systems are developed with a keen understanding of human behavior, tendencies, and limitations. By grounding AVP system design in principles of human-centered design, we can promote safety, enhance user satisfaction, and ensure a harmonious interaction between the driver and the automated system.

Incorporating human factors principles into AVP systems also involves anticipating and designing for the diverse responses of users. This could mean tailoring systems to different levels of driving expertise, stress tolerance, and trust propensity, thus accommodating a wider range of users. As AVP technologies advance, it is crucial that the systems are designed not just for optimal technical performance, but also for the complexities of human interaction, ensuring that the technology is not only used but trusted and understood by its users.

4.3 Limitations and Future Research

The study, while shedding light on the influence of AVP systems on cognitive workload and trust, does have its limitations. The participant demographic was confined to male university students with prior experience with the AVP system, which may not accurately represent the broader population. This limitation raises questions about the generalizability of the findings. Future research should aim to include a more diverse participant pool in terms of gender, age, driving experience, and familiarity with AVP technology to ensure a broader applicability of the results.

Another limitation lies in the controlled experimental setup, which, despite its real-world context, cannot fully replicate the unpredictability and complexity of on-road driving conditions. Subsequent studies should attempt to assess the AVP system's impact on workload and trust in a variety of real-world scenarios, including different traffic conditions, parking environments, and during system malfunctions or failures. Understanding how drivers respond to AVP system errors is crucial for developing robust systems that can handle unexpected situations and maintain driver trust.

Moreover, the relatively small sample size could have affected the statistical power of the study, potentially contributing to the absence of significant results in some measures. Future studies with larger sample sizes are needed to validate and expand upon the observed trends.

In addition to a more diverse and extensive sample, longitudinal studies could provide insights into how trust and workload evolve over time with prolonged use of AVP systems. Such longitudinal data could inform the design of AVP systems that adapt to individual user's changing levels of trust and workload over time, offering a personalized user experience.

5 Conclusion

This study's investigation into the impacts of AVP systems on driver workload and trust provides empirical evidence supporting the potential of AVP technology to enhance the parking experience. The data suggest a trend towards reduced driver stress and an increase in trust, indicating that AVP systems may offer substantial benefits in urban mobility by alleviating some of the mental demands associated with manual parking. While the findings regarding eye-tracking metrics, face orientation, and HRV indicate positive outcomes, the increased mobile device usage during AVP underscores the complexity of the human-automation interaction and the potential for overreliance on technology. The study highlights the critical need for a human-centered approach in the design and implementation of automated driving technologies to ensure safety, reliability, and user acceptance.

Acknowledgments. This work was supported in part by National Key R&D Program of China (grant number 2022YFB2503405), in part by the Tsinghua University-Toyota Joint Research Center for AI Technology of Automated Vehicle (grant number TTAD2023-05), in part by Science and Technology Innovation Key R&D Program of Chongqing (grant number CSTB2023TIAD-STX0027), in part by the Natural Science Foundation of Xinjiang Uygur Autonomous Region (grant number 2023D01A53), and in part by National Nature Science Foundation of China (grant number 51965055).

Disclosure of Interests. The authors have no competing interests to declare that are relevant to the content of this article.

References

1. Banzhaf, H., Nienhüser, D., Knoop, S., et al.: The future of parking: a survey on automated valet parking with an outlook on high density parking. In: 2017 IEEE Intelligent Vehicles Symposium (IV), pp. 1827–1834. IEEE (2017)
2. Huang, C., Lu, R., Lin, X., et al.: Secure automated valet parking: a privacy-preserving reservation scheme for autonomous vehicles. IEEE Trans. Veh. Technol. **67**(11), 11169–11180 (2018)
3. Khalid, M., Wang, K., Aslam, N., et al.: From smart parking towards autonomous valet parking: a survey, challenges and future Works. J. Netw. Comput. Appl. **175**, 102935 (2021)
4. Timpner, J., Friedrichs, S., Van Balen, J., et al.: k-Stacks: High-density valet parking for automated vehicles. In: 2015 IEEE Intelligent Vehicles Symposium (IV), pp. 895–900. IEEE (2015)
5. Ma, J., Feng, X., Gong, Z., et al.: Creating appropriate trust in automated valet parking system. J. Phys. Conf. Ser.. IOP Publishing, **1549**(5), 052059 (2020)
6. Li, Q., Wang, Z., Wang, W., et al.: A human-centered comprehensive measure of take-over performance based on multiple objective metrics. IEEE Trans. Intell. Transp. Syst. **24**(4), 4235–4250 (2023)
7. Min, K.W., Choi, J.D.: Design and implementation of autonomous vehicle valet parking system. In: 16th International IEEE Conference on Intelligent Transportation Systems (ITSC 2013), pp. 2082–2087. IEEE (2013)

8. Hadi, A.M., Li, Q., Wang, W., et al.: Influence of passive fatigue and take-over request lead time on drivers' take-over performance. In: Advances in Human Aspects of Transportation: Proceedings of the AHFE 2020 Virtual Conference on Human Aspects of Transportation, July 16–20, 2020, USA, pp. 253–259. Springer (2020)
9. Meng, T., Huang, J., Chew, C.M., et al.: Configuration and Design Schemes of Environmental Sensing and Vehicle Computing Systems for Automated Driving: A Review. IEEE Sensors Journal (2023). [Note: Year of publication and volume/issue number needed]
10. Stapel, J., Mullakkal-Babu, F.A., Happee, R.: Automated driving reduces perceived workload, but monitoring causes higher workload than manual driving. Transport. Res. F: Traffic Psychol. Behav. **60**, 590–605 (2019)
11. Tenhundfeld, N.L., De Visser, E.J., Haring, K.S., et al.: Calibrating trust in automation through familiarity with the autoparking feature of a Tesla Model X. J. Cognitive Eng. Decision Making **13**(4), 279–294 (2019)
12. Li, Q., Su, Y., Wang, W., et al.: Latent hazard notification for highly automated driving: expected safety benefits and driver behavioral adaptation. IEEE Trans. Intell. Transp. Syst. (2023)
13. Tenhundfeld, N.L., de Visser, E.J., Ries, A.J., et al.: Trust and distrust of automated parking in a Tesla Model X. Hum. Factors **62**(2), 194–210 (2020)
14. Barbe, D., Chao, L., Busolin, M.: Human based rating approach for automated valet parking function evaluation. In: 2020 3rd International Conference on Robotics, Control and Automation Engineering (RCAE), pp. 95–102. IEEE (2020)
15. Esen, H., Kneissl, M., Molin, A., et al.: Validation of automated valet parking. In: Validation and Verification of Automated Systems: Results of the ENABLE-S3 Project, pp. 207–220 (2020)
16. Tomzcak, K., Pelter, A., Gutierrez, C., et al.: Let Tesla park your Tesla: Driver trust in a semi-automated car. In: 2019 Systems and Information Engineering Design Symposium (SIEDS), pp. 1–6. IEEE (2019)
17. Li, Q., Naumenko, A., Fang, Q., et al.: Influence of the relative position of surrounding traffic on drivers' take-over performance. In: International Conference on Applied Human Factors and Ergonomics, pp. 403–409. Springer (2021)
18. Schwesinger, U., Bürki, M., Timpner, J., et al.: Automated valet parking and charging for e-mobility. In: 2016 IEEE Intelligent Vehicles Symposium (IV), pp. 157–164. IEEE (2016)
19. Heikoop, D.D., de Winter, J.C.F., van Arem, B., et al.: Acclimatizing to automation: driver workload and stress during partially automated car following in real traffic. Transport. Res. F: Traffic Psychol. Behav. **65**, 503–517 (2019)
20. Mehler, B., Reimer, B., Coughlin, J.F., et al.: Impact of incremental increases in cognitive workload on physiological arousal and performance in young adult drivers. Transp. Res. Rec. **2138**(1), 6–12 (2009)
21. Holthausen, B.E., Wintersberger, P., Walker, B.N., Riener, A.: Situational trust scale for automated driving (STS-AD): development and initial validation. In: 12th International Conference on Automotive User Interfaces and Interactive Vehicular Applications, pp. 40–47 (2020)

Verification of Continuous Recovery Operation Using Teleoperation System for Autonomous Vehicles - Consideration Focusing on Impact of Road Markings

Tomonori Yasui[1,2](✉), Kouyou Otsu[2], and Tomoko Izumi[2]

[1] Mitsubishi Electric, 8-1-1 Tsukaguchi-honmachi, Amagasaki-shi, Hyogo 661-8661, Japan
Yasui.Tomonori@ea.MitsubishiElectric.co.jp

[2] Ritsumeikan University, 1-1-1 Noji-higashi, Kusatsu 525-8577, Shiga, Japan
{k-otsu,izumi-t}@fc.ritsumei.ac.jp

Abstract. The concept of remote-type autonomous driving was proposed to complement the imperfections of autonomous driving systems, where human operators intervene remotely to ensure proper driving. In a remote-type automatic driving scheme, it is desirable for an operator to monitor multiple vehicles simultaneously to increase work efficiency. To realize this operational structure, this study examines the cognitive challenges faced by operators when multiple recovery requests are simultaneously received from multiple autonomous vehicles. We conducted a user study using a driving simulator by simulating a situation in which recovery requests were continuously received from 10 autonomous vehicles to a single operator. From the results, it is found that the working time increased when there were a series of requests for intervention on factors related to road markings. We then conducted an additional experiment to verify the impact of road markings in a situation requiring recovery. To test this effect in detail, a test similar to the first experiment was conducted in a more realistic scenario involving several traffic participants. Although no statistical differences are identified, the mean work time increases slightly when there are road markings in consecutive situations.

Keywords: Autonomous vehicle · Teleoperation system · Human error · Efficiency

1 Introduction

Autonomous driving systems are expected to be widely used to improve the convenience and efficiency of transportation. In our study, we focus on a challenge in the context of Level 4 autonomous driving, as defined by SAE International [1]. In such scenarios, autonomous driving systems enable system-controlled driving within a predefined operational area called the Operational Design Domain (ODD). However, if an autonomous vehicle encounters a situation that deviates from these predefined conditions owing to certain events, the system cannot continue driving autonomously.

H. Krömker (Ed.): HCII 2024, LNCS 14732, pp. 211–229, 2024.
https://doi.org/10.1007/978-3-031-60477-5_16

One method for safely and seamlessly implementing autonomous driving systems is "remote-type autonomous driving". In remote-type autonomous driving, a human operator monitors and manages vehicles from a distant location using a teleoperation system. This system allows remote operators to intervene in autonomous vehicles according to the circumstances surrounding them. The system was designed to allocate multiple vehicles to a single operator to reduce human costs. For example, according to the roadmap outlined by the Ministry of Economy, Trade, and Industry in Japan [2], there are areas that have set the goal of increasing the number of autonomous vehicles by remote monitoring. Therefore, it is necessary to develop a teleoperation system capable of handling a large number of vehicles simultaneously. However, there are some challenges in realizing such operations in remote-type autonomous driving systems.

A significant challenge in this operational setup is the workload imposed on operators. In the context of Level 4 autonomous driving, there is no obligation for human monitoring as long as the autonomous driving system remains within its ODD. Therefore, an operator is required to intervene only when the autonomous driving system is unable to continue driving. Therefore, when human intervention or remote assistance is needed, the teleoperation system notifies the operator, and the operator must assess the vehicle's situation accurately within a short timeframe. In addition, when replacing traditional route-based buses with a specific autonomous driving system, multiple autonomous buses may operate on the same route or in the same area, which implies that the road environment and surroundings will be similar for each bus. In this scenario, when an event that prevents autonomous driving occurs on one bus, there is a heightened risk that other buses will also encounter factors causing a departure from the conditions for autonomous driving because they all traverse similar situations. Contributing factors include weather conditions, surrounding congestion, and road characteristics. This means that operators need to recognize various similar situations consecutively and immediately and handle them appropriately. This consecutive recovery operation in similar situations may be a critical issue in realizing the assignment of a greater number of vehicles to one operator.

In this study, we aim to investigate the challenges in the operator recovery process for remote-type autonomous driving. In particular, our study focuses on issues of how an operator understands a situation of each vehicle when dealing with many vehicles continuously using a teleoperation system. Specifically, we employ a driving simulator to replicate situations in which multiple autonomous vehicles in a similar environment cannot continue driving for various reasons. We then examine how dealing with similar types of requests consecutively affects the workload and errors of the operator. In our experiment, participants were tasked with a continuous recovery operation using a prototype that resembled existing teleoperation systems. This study resents and discusses the results of the two experiments: In the first experiment, we compared the difference in the working time between performing similar and different tasks in a continuous recovery operation situation. In most cases, no significant increase in work time is observed; however, a significant increase in working time is observed in the case of a request caused by a problem associated with road markings. Based on these results, an additional experiment was conducted to verify the effect of road markings on continuous recovery operations. To investigate whether the results of the previous experiment are due to specific factors in the mark, we also treat a different request situation owing to the error in the line trace caused by the blurring of roadside lines. Moreover, the experiment asked participants to perform recovery operations in congested situations compared with the first experiment, as the low-traffic situation in the first experiment was considered unsuitable for testing the assumed hypotheses of this study.

2 Related Studies

2.1 Previously Identified Problem of Remote Operation Systems

A frequently encountered issue in remote autonomous driving systems is the degradation of operational efficiency arising from communication delays. Previous studies proposed new user interfaces that consider the operational challenges associated with communication delays [3, 4]. These studies focused on the operational aspects of driving. However, it is imperative to extend our focus to teleoperation systems that encompass the intricacies of situational awareness and decision-making processes. As mentioned in the preceding section, teleoperation systems are expected to promptly comprehend a vehicle's status and handle scenarios that require continuous remote vehicle operation. Consequently, it is important not only to tackle the issues of teleoperating the vehicle but also to provide support for perceiving the vehicle's situation and making decisions to plan appropriate operations.

2.2 Remote-Type Autonomous Vehicle System for Multiple Vehicles

Ding et al. [5] proposed a method for actively managing the operation of multiple autonomous vehicles to distribute the timing at which vehicles request monitoring from operators. In this method, environments such as poorly visible intersections or road construction that may hinder the operation of an autonomous driving system are input into the system in advance. The operation is then adjusted to distribute the time at which the vehicles pass through these environments. This helps reduce the risk of congestion in requests for the monitoring and operation of vehicles, thereby enabling the operation of multiple autonomous vehicles with a small number of operations. However, because there are many dynamically changing factors in traffic conditions, it is difficult to predict the conditions that may hinder the operation of autonomous driving systems. Therefore, teleoperation systems must be capable of handling requests even in congested situations, thereby providing a robust solution.

Furthermore, Yokota et al. [6] verified the appropriate amount of information for operators to effectively recover two autonomous vehicles that simultaneously could not continue system-controlled driving due to on-street parking situations. It is critical to investigate the limitations on the amount of information that a signal operator can process simultaneously. However, the verification was specific to situations involving street parking. To develop a teleoperation system that can provide support under various situations, it is essential to conduct similar verifications across diverse scenarios.

3 Our Hypotheses

In this study, we discuss the challenges in situations where an intervention request is made from one vehicle immediately after completing a recovery operation of another vehicle during remote-type autonomous driving. In this case, an operator must continuously understand the reason for the intervention and the situation of the vehicles. We propose two hypotheses regarding the challenges that arise in such situations. This section describes the remote-type autonomous driving targeted in this study and the two hypotheses.

3.1 Assumed Teleoperation System in This Study.

We assume a level of driving automation is "Level 4" defined by SAE International and a single operator operates 10 vehicles. This study considers the workload of the operator in performing recovery operations, such that when the autonomous driving system of a vehicle cannot continue driving owing to any event, the vehicle is restored to a drivable state.

Referring to previous remote-type autonomous driving systems, we assume the recovery operation procedure in this study is as follows: First, the autonomous driving system of a vehicle requires a recovery operation to a remote operator because it cannot continue to drive. Then, the operator confirms information (e.g.: footage of on-board camera from the vehicle) to estimate the factor the autonomous vehicle cannot continue driving (Henceforth, the factor is called "intervention factor"). Finally, the operator remotely operates the vehicle to recover it. This procedure allows the operator to perform recovery operations on a single vehicle. However, in this study, we consider a situation in which some vehicles may simultaneously request recovery operations. In this case, the operator repeats the procedure until no vehicles request a recovery operation.

3.2 Assumed Hypothesis About Human Error

The first hypothesis is that human errors are induced in specific situations. Specifically, this is "if something that an operator focused on during a recovery operation of one vehicle appears in a surrounding environment the next vehicle, the common something may trigger human error." In the case of the sequential recovery of multiple vehicles, it is possible for an operator to encounter similar situations. In this case, the repeated appearance of an object to which the operator's attention was directed during the previous recovery of a vehicle may have influenced the operator's decision-making. Consequently, if the request from the second vehicle involves the object to which the operator paid attention during the first recovery of the vehicle, there may be an increased risk that the operator overlooks or misjudges other factors to be considered. That is, if an object that an operator focuses on during a recovery operation of one vehicle appears in an environment surrounding the next vehicle, the common object may trigger human errors.

3.3 Assumed Hypothesis About Decreased Work Efficiency.

The second hypothesis is that continuous recovery operations reduce work efficiency. During consecutive recovery operations, it is difficult for operators to immediately grasp situations of vehicles remotely. As mentioned in the previous section, there is a potential risk of confusion and error during the consecutive recovery operations. In situations in which operators are prone to confusion, a more cautious approach to situational awareness and judgment is required. This means that, compared to a recovery operation with a single vehicle, there is a potential for reduced efficiency in the process of operators recognizing a situation of each vehicle in consecutive recovery operations with multiple vehicles.

4 Experiment

4.1 Experimental Purpose and Method

An experiment was conducted to validate the hypotheses presented in the previous section. The experiment utilized a driving simulator, Prescan [7], provided by Siemens, to simulate scenarios in which autonomous vehicles simultaneously stopped owing to different causes. The participants were asked to perform consecutive recovery operations using the prototype system described later.

In this experiment, we designed ten scenarios. In each scenario, an intervention request for a different reason notifies an operator through the system. Table 1 summarizes these scenarios. We set the five factors to attract an operator's attention during recovery tasks, and the two scenarios are set for each factor. The two scenarios with the common factors have the following characteristics: one is that the common factor itself triggers the intervention request, and the other is that the common factor appears but does not trigger the intervention request. The intervention factors for the scenarios are configured based on the reports from previous autonomous driving experiments conducted in Japan.

In this experiment, we focus on the similarity between the scenarios experienced by the participants and those they experienced just before to verify the hypotheses regarding consecutive recovery operations mentioned in Sect. 3. In particular, we investigate whether differences in human errors and work efficiency are observed in the consecutive recovery operation depending on whether the participants experience a paired scenario with the same factors immediately before. To verify the influence of the common factors, the scenarios do not include other traffic-related objects, such as traffic participants and signage, other than those set as interventions or common factors.

4.2 Prototype System for This Experiment

We implemented a prototype teleoperation system that participants used for recovery operations. The system was developed with reference to those used in the autonomous driving demonstration experiments [8–12]. This section presents an overview of the proposed system.

Figure 1 shows the appearance and screens of the prototype system. The system consists of two displays and a steering-wheel controller. One display (Fig. 1(b)) primarily shows real-time footage from the onboard camera of a vehicle. The forward-facing camera image is displayed at the center of the screen, while the rear-, right-, and left-facing camera images are shown at the top of the screen. In addition, essential information for vehicle operations, such as speed, is displayed at the bottom of the screen.

Table 1. The ten types of vehicle recovery scenarios for this experiment.

No	Common factor	Abstract of scenario	Correct recovery operation
1	Pedestrian	The vehicle cannot start moving because pedestrians stop in front of the vehicle while it was stationary	Drive manually to avoid the pedestrians
2		The communication speed is decreasing There is a pedestrian at a certain distance from the vehicle	Contact maintenance personnel
3	Oncoming car	Within a narrow lane, an oncoming vehicle is approaching. It is necessary to cross the lane to pass the oncoming vehicle, however, the autonomous driving system cannot make the decision to over the lane	Drive manually to pass the oncoming car
4		The vehicle stops due to a malfunction of the on-board sensors An oncoming car is present in the forward direction of the vehicle, however, as it is an opposite lane road, the oncoming car does not affect the vehicle's travel	Contact maintenance personnel
5	Traffic sign	The stop traffic sign is obscured by branches and grass, so the vehicle cannot detect the sign's content and then it stops	Drive manually to pass the sign
6		The vehicle pulls over the shoulder to yield the right of way to an approaching emergency vehicle from behind There is a clearly visible and recognizable road sign ahead of the vehicle	Resume autonomous driving

(continued)

Table 1. *(continued)*

No	Common factor	Abstract of scenario	Correct recovery operation
7	Car facing the same direction	The autonomous driving system stops as it cannot overtake due to the presence of a parked car ahead on the roadway	Drive manually to pass the parked car
8		The vehicle detects vibrations from running over a small stone on the road, prompting an abnormality assessment and resulting in a stop A car is parked completely within the shoulder ahead in the same lane as the vehicle	Resume autonomous driving
9	Traffic light	Approaching a signalized intersection including crosswalks, the vehicle is unable to detect the traffic light colors using the camera due to backlighting and stops	Drive manually to pass the traffic light
10		In front of the vehicle, there is a road marking indicating a no-stopping zone (this is a Japanese road marking that looks like a zebra zone), causing a failure in lane detection and resulting in a stop Beyond the no-stopping zone, there is a signal unaffected by backlight and so on	Drive manually to pass the road markings

The other display shows the three main sections. The first section displays the status of the onboard equipment and the information detected by the autonomous driving system. In each case of normal operation, it displays "Normal" in green, and in the event of any abnormal condition, it indicates "Abnormal" in red. The second section includes the three buttons for the input required during recovery operations. Two of these buttons are positioned to resume autonomous driving and manual driving by operators. The other button is used to contact and dispatch maintenance personnel to the vehicle site in situations where intervention factors are not resolved remotely. These buttons are selected or pressed through button operations on the steering wheel controller. The last section displays positional information on the vehicle. On the map, a red marker is drawn at the current location of the vehicle and the planned route is depicted in light blue.

The participants are able to comprehend the status of the vehicle from these screens. They operate the vehicle using a steering wheel controller and conduct driving operations similar to actual driving. The procedure for the recovery operation using this system is as follows.

Step 1. The participant receives a notification of an intervention request from a vehicle in which autonomous driving became challenging.
Step 2. Upon confirming the notification, the participant presses the button to accept the request.
Step 3. The participant estimates the vehicle's condition and the intervention factor from the information displayed on the screens and then decides on a proper recovery operation.
Step 4. The participant presses one of the buttons: "Autonomous driving," "Manual driving," or "Contact maintenance personnel," based on his/her decision.
Step 5. If the participant presses the "Manual driving," the steering wheel controller is available to drive the vehicle manually. To end the manual driving, the participant presses the "Autonomous driving" button.
Step 6. The participant completes the recovery operations.
Step 7. Steps 1–6 are repeated until there is no more intervention request from any vehicle.

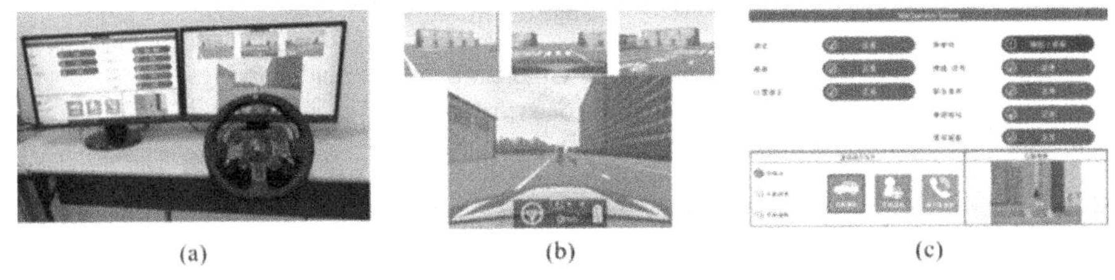

Fig. 1. (a) The prototype system for this experiment. (b) The display for driving. (c) The display for vehicle information.

4.3 Experimental Procedure

First, the participants received the explanations about an overview of the autonomous driving system and the role of the operator in remote-type autonomous driving. Next, the participants were instructed on the features of automated vehicles and provided the overview of remote-type autonomous driving in advance. In addition, they were informed of the reasons why intervention requests were made to the operator in remote-type autonomous driving and how the operator should handle each request. Figure 2 illustrates the anticipated responses explained to the participants. They were then instructed on how to use the prototype. Especially, they were instructed to press the corresponding button immediately upon deciding how to respond to the request. This is because it measures the time that they spent on situational awareness and deciding on a recovery plan. After these instructions, we obtained consent from the participants to participate in the experiment, and the participants practiced recovery operations with three training scenarios to become familiar with the system and steering wheel controller. After a break, they performed consecutive recovery operations for the ten intervention scenarios listed in Table 1 as the main task.

To avoid order effects, the order of the scenarios was changed by participants. This order was set to ensure equal distribution of participants between those who had a common factor appearing consecutively and those who did not. For example, in the case of Scenario 1, the number of participants who participated in Scenario 2, including the common factor with Scenario 1, immediately prior was adjusted to approximately equal the number of participants who did not participate in Scenario 2 immediately prior. Furthermore, during the experiment, the participants wore the eye tracker Tobii Pro Glasses 3 [13], to measure their gaze points. After the experiment, we reviewed the videos recorded from the eye tracker with the participants and freely articulated their observations and thoughts during the recovery operations for each scenario. Subsequently, the participants were asked the following questions for each scenario:

1. What did you estimate as the intervention factor?
2. Why did you estimate that it was a particular intervention factor?
3. What kind of recovery operation did you consider appropriate?
4. Why did you think that your recovery operation was appropriate?

This experiment and subsequent experiments were conducted with the approval of the Ritsumeikan University Ethics Committee (approval number: Kinugasa-Hito-2022-68).

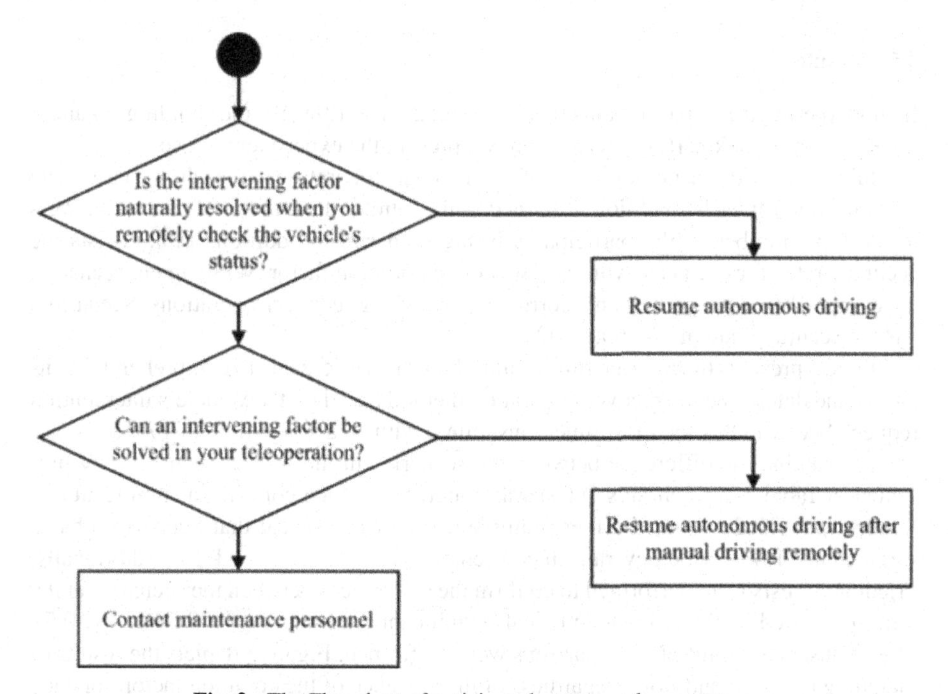

Fig. 2. The Flowchart of anticipated responses by operators.

4.4 Evaluation Method

The hypothesis regarding human error is evaluated based on the accuracy rate of the participants' recovery operations in each scenario. Each scenario is assumed to have a predefined correct recovery operation, as listed in Table 1. The accuracy rate is calculated based on whether the participant's recovery operation is the same as the assumed correct recovery operation. We compare the accuracy rates between the case where the two scenarios involving the common factor are consecutive and the case where the prior scenario did not have the common factor.

The assumed hypothesis related to work efficiency is evaluated based on the time required for the participants to comprehend the situation and determine the recovery operation method from the vehicle's intervention request. This time is measured starting when the participant accepted the notification through the prototype system operation in Step 2 of Sect. 4.2, and ending when the button was pressed in Step 4 of Sect. 4.2. Similar to the accuracy rate, the analysis compares the two cases in which the immediately preceding scenario involved the common factor. Henceforth, the case where the immediately preceding scenario involves a common factor is called "Case with a common factor," and the case where it does not is called "Case without a common factor."

Using the aforementioned evaluation criteria, we compare the differences between each scenario to analyze their characteristics. If the hypotheses are confirmed, we discuss the results based on the interview content in our analysis. In situations where objective data focusing on specific objects are deemed necessary, we use the gaze-point results measured by an eye tracker.

4.5 Results

In this experiment, 19 participants (average age: 22.4 ± 1.9 years old) holding Japanese driver's license took part. In this section, we present the experimental results.

Table 2 lists the accuracy rates of the recovery operations in each scenario. The accuracy rate for each condition is defined as the ratio of the number of correct answers to the total number of the participants in the scenario. In addition, Table 2 lists the accuracy rates for each case with and without a common factor. While many scenarios have over 70% of the participants correctly performing recovery operations, Scenario 8 has the accuracy rate of less than 30%.

Table 3 presents the average time required for the participants to comprehend the situation and determine the recovery operation method based on the vehicle's intervention request. Scenario 8 is the most time-consuming, with an average time of 17.0 s.

To examine the difference between the scenarios in the overall accuracy rate presented in Table 2, a Cochran's Q test was conducted, which confirms a significant difference ($p = .002 < .05$). Further multiple comparisons reveal that Scenario 8 has a significantly lower accuracy rate than Scenarios 1, 2, 3, 4, 7, and 10. Additionally, Friedman's tests were performed to confirm the differences between the scenarios at the times presented in Table 3, which reveals significant differences ($p = 8.58\ e^{-9} < 0.5$). Then, subsequent multiple comparisons were performed. Figure 3 depicts the results of including the two conditions, regardless of the presence of the common factor, for each scenario. In terms of the time required for situational awareness and judgment, Scenario

8 takes significantly longer than Scenarios 1, 3, 5, 7, and 10. Furthermore, Scenario 2 requires more time than Scenarios 5 and 7.

Table 2. Accuracy rates of recovery operations in each scenario.

Scenario no	Cases with a common factor	Cases without a common factor	Entirety
1	100% (10/10)	89% (8/9)	95%
2	89% (8/9)	100% (10/10)	95%
3	89% (8/9)	100% (10/10)	95%
4	100% (10/10)	100% (9/9)	100%
5	60% (6/10)	89% (8/9)	74%
6	78% (7/9)	60% (6/10)	68%
7	100% (9/9)	90% (9/10)	95%
8	30% (3/10)	22% (2/9)	26%
9	80% (8/10)	89% (8/9)	84%
10	100% (9/9)	90% (9/10)	95%

Table 3. Average time required to understand the situation and decide on recovery operation method in each scenario (sec.).

Scenario no	Cases with a common factor	Cases without a common factor	Entirety
1	8.8	10.8	9.8
2	13.2	15.7	14.5
3	9.3	9.2	9.2
4	8.2	12.8	10.4
5	10.1	9.0	9.6
6	11.9	12.6	12.3
7	7.3	8.8	8.1
8	20.0	13.6	17.0
9	15.9	9.2	12.7
10	8.2	10.0	9.1

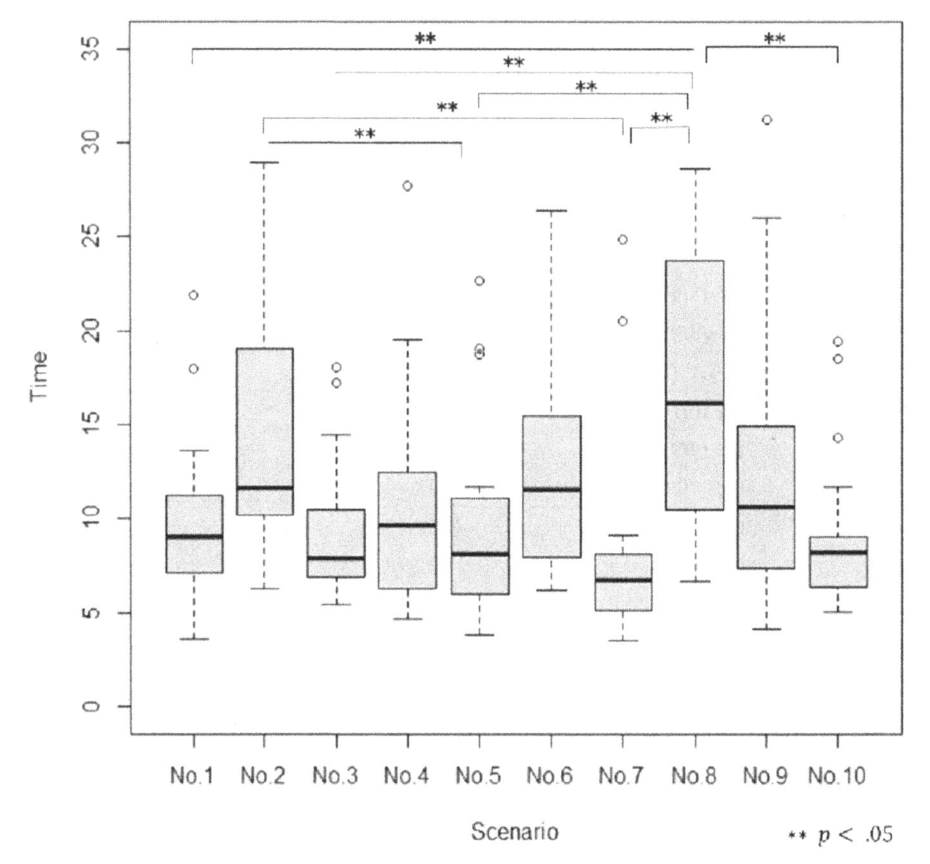

Fig. 3. Distribution of time required to understand the situation and decide recovery operation method for each scenario.

4.6 Discussions

First, the characteristics of each scenario are discussed. These results indicate the difficulty of the recovery operation in Scenario 8. In Scenario 8, a small stone is observed in the area circled in red in Fig. 4 (behind the vehicle). During the recovery operation process, the operator must recognize this intervention factor based on the notification of abnormal vibrations and camera views displayed on the prototype system. However, none of the participants mentioned noticing the small stones in the interview after the experiment. Additionally, eye-tracking data reveals that none of them fixates on the location of a small stone. The system has the rear-facing camera image displayed at approximately one-third of the size of the forward-facing image, suggesting that the presence of intervention factors in small-display camera images may often go unnoticed.

Next, we discuss the hypothesis of human error in consecutive recovery operations. To verify this, we examine the differences in the accuracy rates between the two cases with and without a common factor. Fisher's exact probability test was conducted for each scenario using the results presented in Table 2. No significant difference is observed between the cases in any scenario. Therefore, the hypothesis related to human error is not supported in this experiment. This is attributed to the simplified traffic environment simulated in the driving simulator, which might have reduced the task difficulty. It can be considered that if the experiment is conducted in a more realistic traffic environment, such as introducing more traffic participants, the hypothesis could be supported.

Finally, to verify the hypothesis of reduced work efficiency, we examine the differences in the time it took for participants to comprehend the situation and determine the recovery operation method from the vehicle's intervention request between the two cases. Shapiro-Wilk tests for data normality and F-tests for data homogeneity were conducted for each scenario. Following confirmation of normality and homogeneity of variance (two-tailed: $p > = .05$), Student's t-tests were employed for Scenario 1 and 8, Welch's t-test for Scenario 4 and 9, and Mann-Whitney U tests were employed for other scenarios where normality and homogeneity of variance were confirmed. Only Scenario 9 shows a significant difference ($p = .034 < .05$) between the two cases, where Scenario 10 was performed immediately before and where it was not. In Scenario 9, the average time it took for participants to comprehend the situation and determine the recovery operation method increases by 6.7 s in the case with the common factor compared with the case without the common factor. However, in the interviews, none of the participants reported observing the traffic light during the recovery operation of Scenario 10, which is the common factor between Scenarios 9 and 10. Therefore, even if the participants observed the traffic light during Scenario 9, it would be difficult to argue that the traffic light affected their work efficiency. In other words, although the traffic light does not impact the work efficiency, another element introduced in Scenario 10 may have influenced the recovery operation of Scenario 9. This possible element is unique to Scenario 10, does not appear in other scenarios, and can be considered a common element in Scenario 9.

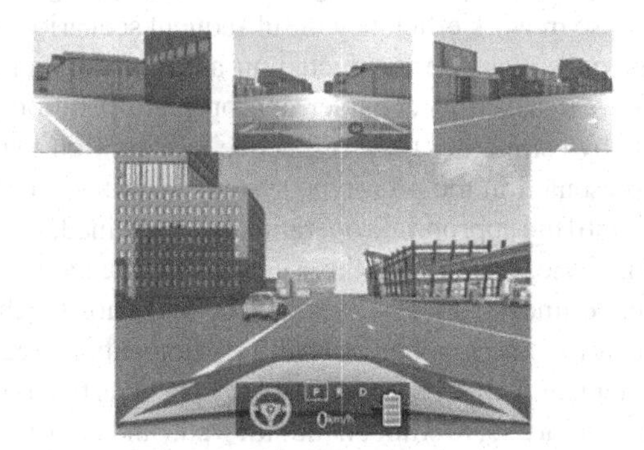

Fig. 4. Image of the onboard camera in Scenario 8 (the red circle shows the position of the small rock.).

We look for other factors common to Scenarios 9 and 10, and the road markings on the street fulfill the above conditions. In Scenario 9, the road markings for a pedestrian crossing in front of the vehicle, and the no-stopping zone are set as the intervention factor in Scenario 10. Hence, it is possible that road markings function as the common factor without being intended for in advance. To explore this further, we confirm the eye-tracking data from 17 participants for whom gaze data are available to calculate the time participants spent focusing on the pedestrian crossing during Scenario 9 until they performed the recovery operation. The results indicate that the participants who performed Scenario 10 just before spent 1919.0 ms focusing on the pedestrian crossing on average, while those who did not spent 679.9 ms. That is, it is observed that when Scenario 10 is conducted immediately before, the participants tend to see the road markings for approximately one second longer than the case when it is not. As a result, when remotely handling a vehicle with road markings around it, the recovery operation time may vary between cases with and without a common factor.

5 Additional Experiment

5.1 Purpose and Scenarios for Recovery Operation

In the experiment described in the previous section, it is confirmed that the work efficiency may decrease when recovering from situations where distinctive road markings appear immediately after recovering from scenarios where road markings indicating a no-stopping zone are intervention factors. However, in that experiment, road markings were not intentionally set as the common factor that confounded operators' recognition. In addition, as mentioned in the previous section, we discussed the inability to verify the hypothesis regarding human errors because the previous experiment was conducted in the simple situations with few traffic participants. Therefore, there are two new questions: First, what is the impact of increased traffic on work efficiency and human error in consecutive scenarios involving road markings? It is possible that increasing traffic leads to different tendencies in situational perceptions. The second question is whether a scenario that includes different road markings from a no-stopping zone, such as roadside lines, leads to a decrease in work efficiency in subsequent scenarios with road markings.

To address these questions, we conducted an additional experiment to verify our results. Participants were asked to consider appropriate operations in situations with consecutive requests by estimating their respective intervention factors with reference to the information presented in the prototype system. However, in this experiment, we consider the process until the appropriate operation is determined, which does not include the driving operation phase. In addition, the scenarios for the recovery tasks were recreated for additional experiments. Table 4 presents the scenarios for the additional experiments. In these scenarios, the number of traffic participants increases compared with the previous experiment, and all scenarios are set to appear as two pedestrians and many vehicles. The pedestrians are set to walk on the sidewalk, and the vehicles are set to travel continuously in the oncoming lane at a speed of approximately 30 km/h with an interval of approximately 50 m. In Scenarios 1 and 2, as in the previous experiment, distinctive road markings, a no-stopping zone, and a pedestrian crosswalk are set as the common factor. However, because Scenarios 1 and 2 in the additional experiment have many traffic participants, they differ from Scenarios 9 and 10 in the previous experiment in terms of environmental complexity. Scenarios 3 and 4 in Table 4 have roadside lines that are different road markings from a no-stopping zone or pedestrian crosswalk. The purpose of this additional experiment is to verify whether these common factors would lead to situational confusion and result in human errors or decreased work efficiency when the scenarios with the common factors appear consecutively in recovery operations. In particular, we examine whether the results differ in trend from the results of Scenario 9 after applying Scenario 10 in the previous experiment. In addition, four other scenarios are added as Scenario 5 through 8. These scenarios do not include the factors related to road markings and are included to simulate realistic consecutive recovery operations.

5.2 Evaluation Method

As in the previous experiment, certain scenarios were designed using pairs that shared the common factors. Similar to the previous experiment, this experiment aims to compare the cases with and without a common factor in contentious recovery operations to investigate the impact it on work efficiency and human errors. Specifically, for the results of Scenario 2, we examine the difference between the case of recovering from Scenario 1 just before and that of recovering from another scenario. Scenario 1 and 2 have a no-stop zone as the common factor. Similarly, for the results of Scenario 4, we examine the difference between the case of recovering from Scenario 3 just before and that of recovering from another scenario. Scenario 3 and 4 have roadside lines as the common factor.

The evaluation procedure was nearly the same as that for continuous recovery operations in the previous experiment. However, in the additional experiment, the participants repeated steps 1 through 4, explained in Sect. 4.2, without steering wheel operation during the recovery process. However, the order of scenarios for recovery operations was determined by considering the following points. The first point is to ensure that the number of participants doing the case with a common factor is equal to the number of participants doing the case without a common factor. This restriction is the same as that used in the previous experiment. The second point concerns the common factor, roadside lines, in Scenarios 3 and 4. If the hypothesis that common factors influence work efficiency is correct, recovery operations immediately after Scenario 3 would be influenced by the roadside lines. Because roadside lines appear in all scenarios in Table 4, based on the hypothesis, Scenario 3 affects the recovery operation of the other scenarios. For example, if Scenario 2 is handled immediately after Scenario 3, it becomes unclear whether the operation in Scenario 2 is influenced by road markings or roadside lines. To avoid this, Scenarios 3 and 4 were handled last in the recovery operation sequence. The order within this pair changed among the participants. When Scenario 3 is implemented before Scenario 4, it is verified as a situation influenced by their common factor; when Scenario 3 is implemented later (i.e., at the end of the sequence of recovery operations), it is verified as a situation not influenced by the common factor.

Table 4. Eight scenarios of vehicle recovery for the additional experiment.

No	Common factor	Abstract of scenario	Correct recovery operation
1	Road markings	In front of the vehicle, there is a road marking indicating a no-stopping zone, causing a failure in lane detection and resulting in a stop Beyond the no-stopping zone, there is a signal unaffected by backlight and so on	Drive manually to pass the road markings
2		The stop traffic sign is obscured by branches and grass, leading to the inability to detect the sign's content and resulting in a halt There is a pedestrian crosswalk in front of the vehicle	Drive manually to pass the sign
3	Roadside lines	The roadside lines are faded, and the detection accuracy has decreased, resulting in stopping	Drive manually to pass the lines
4		The rear camera has a scratch, causing a decrease in detection accuracy and resulting in a stop There are normal roadside lines	Contact maintenance personnel
5		The vehicle cannot start moving because pedestrians stop in front of the vehicle while it is stationary	Drive manually to avoid pedestrians
6		The communication speed is decreasing	Contact maintenance personnel
7		The autonomous driving system stops as it cannot overtake due to the presence of a parked car ahead on the roadway	Drive manually to pass a parked car
8		The vehicle pulls over the shoulder to yield the right of way to an approaching emergency vehicle from behind	Resume autonomous driving

5.3 Results

Twenty participants held Japanese driver's license were joined in this experiment (Mean age: 22.35 ± 1.85). The participants included those who had participated in the previous experiment. This is not considered to affect the results significantly because there is no identical scenario between the two experiments.

First, we present the results to validate our hypotheses regarding human error. The accuracy rates for Scenarios 2 and 4 are listed in Table 5. In Scenario 2, the accuracy rate of the recovery operations when Scenario 1 precedes is 90%, whereas the accuracy rate when handling another scenario immediately before is 100%. There is no significant difference in the accuracy rate between the cases with and without a common factor. Similarly, in Scenario 4, the accuracy rate when Scenario 3 preceded is 90%, whereas it is 100% when handling another scenario immediately before. There is no significant difference in the accuracy rates between the two cases.

Next, we present the results that verify our hypothesis of a decrease in work efficiency. As indicated in Table 6, in Scenario 2, the time required for situational understanding and decision, when Scenario 1 is handled immediately prior, averages 12.37 s, whereas it averages 9.95 s when handling another scenario. Statistical analysis using Student's t-test does not reveal a significant difference ($p = .378 > \ = .05$) between these two cases. Similarly, in Scenario 4, the corresponding average time is 11.14 s when Scenario 3 is handled immediately prior and 9.98 s when handling another scenario immediately prior. This difference is not statistically significant according to Student's t-test ($p = .651 > \ = .05$).

Table 5. Accuracy rates of recovery operations for Scenario 2 and 4 in the additional experiment.

Scenario no	Cases with a common factor	Cases without a common factor
2	90%	100%
4	90%	100%

Table 6. Average time required to understand the situation and decide recovery operation method for scenario 2 and 4 in the additional experiment (s).

Scenario no	Cases with a common factor	Cases without a common factor
2	12.37	9.95
4	11.14	9.98

5.4 Discussions

The additional experiment aims to address the effect of road markings on the efficiency and human error of consecutive recovery operations, especially under conditions where the number of traffic participants increases compared to the previous experiment and where roadside lines, which are different road markings from a no-stopping zone, are common factors.

As mentioned above section, the time required for situational understanding and decision of recovery operation in the case with a common factor increases on average by 2.42 s compared with the case without a common factor. Statistical analysis does not reveal a significant difference between these two cases. However, considering the number of participants in this experiment, it is possible that there is a tendency for road markings to exhibit an increase in time compared to the case without markings. Thus, in scenarios with increased traffic participants, the influence of road markings which are included in the immediately preceding scenario on work efficiency is not observed as much as in the previous experiment. The reason why a significant decrease in work efficiency is not observed between these cases in the additional experiment, while it is confirmed in the previous experiment, might be owing to an insufficient number of participants for the experiment. Another possible reason is that the increase in the number of traffic participants in the additional experiments may have accounted for this difference. The difference between Scenario 9 in the previous experiment and Scenario 2 in the subsequent experiment is the number of surrounding traffic participants. With the increase in traffic participants in the additional experiment, there are more objects to attend to confirm the vehicle's situation. As a result, in the additional experiment, there might be fewer opportunities for the participants to consciously focus on the road markings than in the previous experiment. However, additional experiments are necessary to verify this hypothesis.

Next, we discuss the effects of different road markings in no-stopping zones. Scenarios 3 and 4 in the additional experiment have the roadside lines as the common factor. In the experimental results for Scenario 4, the average time required for situational understanding and decision of recovery operation increases by 1.16 s in the case with a common factor, but this difference is not statistically significant as mentioned above. These results are similar to those in Scenario 2. One possible reason for not confirming the significant difference could be the increased number of traffic participants, similar to the discussion in Scenario 2. In this experiment, we assumed that increasing the number of traffic participants would induce cautious recovery operations and reduce work efficiency, especially in confusing situations including a common factor. However, the expected results of this experiment are not confirmed. The reason of this is possibly that the driving simulator does not provide sufficient realism to encourage a cautious recovery. Owing to the lack of realism, the increase in traffic participants may have only diverted their attention from road markings. Considering this, if Scenario 3 was conducted in a situation with fewer traffic participants, it could have influenced the recovery operations in the immediate subsequent scenario. To verify this, a comparative experiment must be conducted with scenarios that differ only in terms of the number of traffic participants.

6 Conclusion

This study aimed to investigate the challenges regarding human error and a decrease of work efficiency of operators in continuous recovery operations, in order to achieve a situation the number of autonomous vehicles assigned to operators is increased in remote-type autonomous driving. First, we hypothesized that in consecutive recovery operations, a common factor such as pedestrians or signage could influence the operator's perception of the situation of a vehicle, leading to confusion, human error, and decreased operational efficiency. To test these hypotheses, we conducted the experiment using the driving simulator to replicate ten autonomous vehicles that required recovery operations by a remote operator. The participants performed continuous recovery operations using the prototype teleoperation system. The results did not reveal human errors induced by most common factors; however, they suggested that operational efficiency might decrease when consecutively recovering from situations in which road markings appear as intervention factors. In the additional experiment, no significant decrease in work efficiency was observed, although there was a tendency toward a decrease in work efficiency on average. Therefore, in continuous recovery operations, it is possible that work efficiency decreases in the case of road markings as a common factor. However, the extent of this decrease differed depending on the presence of traffic participants.

In future research, we plan to continue verifying the hypotheses, considering the number of traffic participants, while maintaining a focus on safety engineering aspects.

References

1. SAE International. https://www.sae.org/blog/sae-j3016-update
2. Ministry of Economy, Trade, and Industry. https://www.meti.go.jp/english/press/2020/pdf/0512_001a.pdf
3. Prakash, J., Vignati, M., Vignarca, D., Sabbioni, E., Cheli, F.: Predictive display with perspective projection of surroundings in vehicle teleoperation to account time-delays. IEEE Trans. Intell. Transp. Syst. **24**(9), 9084–9097. IEEE (2023). https://doi.org/10.1109/TITS.2023.3268756
4. Zhu, Y., Aoyama, T., Hasegawa, Y.: Enhancing the transparency by onomatopoeia for passivity-based time-delayed teleoperation. IEEE Robot. Autom. Lett. **5**(2), 2981–2986. (2020). doi: https://doi.org/10.1109/LRA.2020.2972896
5. Ding, M., Takeuchi, E., Ishiguro, Y., Ninomiya, Y., Kawaguchi, N., Takeda, K.: How to monitor multiple autonomous vehicles remotely with few observers: an active management method. In: 2021 IEEE Intelligent Vehicles Symposium (IV), pp.1168–1173. IEEE, Japan (2021).: https://doi.org/10.1109/IV48863.2021.9575997
6. Yokota, M., Tsutsumi, S., Hayakawa, S., Ikeura, R.: Research on effects and influence by presenting information on priority order and oncoming vehicle to operators for teleoperation of multiple autonomous vehicles. J. Phys. Conf. Ser. **2107**(1), 012012 (2021). https://doi.org/10.1088/1742-6596/2107/1/012012
7. Siemens. https://plm.sw.siemens.com/en-US/simcenter/autonomous-vehiclesolutions/prescan/
8. The National Institute of Advanced Industrial Science and Technology. https://www.meti.go.jp/meti_lib/report/H30FY/000352.pdf (in Japanese)
9. Tokyu Corporation. https://www.tokyu.co.jp/image/news/pdf/20201124-1.pdf (in Japanese)
10. Aichi, Japan. https://www.pref.aichi.jp/uploaded/attachment/414635.pdf (in Japanese)
11. DeNA Co., Ltd. https://dena.com/jp/press/4450/ (in Japanese)
12. BOLDLY Inc. https://www.softbank.jp/drive/set/data/service/img/shared/pdf_catalog_03.pdf (in Japanese)
13. Tobii. https://www.tobii.com/products/eye-trackers/wearables/tobii-pro-glasses-3

Author Index

H. Krömker (Ed.): HCII 2024, LNCS 14732, pp. 231–232, 2024.
https://doi.org/10.1007/978-3-031-60477-5

Printed in the United States
by Baker & Taylor Publisher Services

Printed in the United States
by Baker & Taylor Publisher Services